Weapons, Wars and Economics

Weapons, Wars and Economics

Major General Partap Narain, I.A. (Retd)
M.A. (Cantab) F.I.E.

Manas Publications
New Delhi-110 002 (INDIA)

MANAS PUBLICATIONS
(Publishers, Distributors, Importers & Exporters)
4858, Prahlad Street,
24, Ansari Road, Darya Ganj,
New Delhi - 110 002 (INDIA)
Ph.: 23260783, 23265523 (O); 27131660, 27143052 (R)
Fax: 011-23272766
E-mail: manaspublications@vsnl.com

First Published 2004

ISBN 81-7049-174-6
Rs. 795/-

Typeset at
Manas Publications

Printed in India at
Nice Printing Press
and Published by Mrs Suman Lata for
Manas Publications, 4858, Prahlad Street,
24, Ansari Road, Darya Ganj,
New Delhi - 110 002 (INDIA)

The Ballad of the Hundred Soldiers

I had hundred comrades,
The finest you could find.
They marched away to battle,
But me they left behind.

In Africa, in Flanders,
In Crete and Singapore,
My friends were torn and shattered
In Corporal Hitler's war.

When winter nights are darkening
And winds howl in the trees,
I hear their voices call me,
And they use words like these:

"Comrade O former Comrade
Who left us in the war,
How have you since been living,
These twenty years and more?

"We lie here dead, forgotten,
Life's luckless, lost young men;
But still our voice can haunt you:
What have you done since then?

"Where is that gay Utopia
You vowed to build for our sons:
That world of peace and friendship,
Of justice, not of guns?
"Each day you solely squander
In quest of selfish gain,
Our ghosts will rise and taunt you
With words of bitter pain:

"Raise up that blood stained banner
That bravely once we bore,
And march to fight life's battles
In peace just as in war.

"Or we shall haunt your memory
On winter nights like these
When darkening skies are lowering
And winds howl in the trees."

Alan Moray Williams

Acknowledgements

I thank my friends who have generously helped me to present this book; they are a legion. I am sad to record that today majority of my comrades who went to Hitler's war are no more. The book is a tribute to friends Alan, Sampangi Raj, Hari, Williams, Litty, Yash, Inamdar, Moti Sagar, Bhide and Rahman—who have contributed by their personal accounts, and thereby enhanced the value of this book—by widening the canvas. Some of them are no longer with us to receive your approbation, but I shall send the copies of the book to their families as memento. I only hope that I have recorded them truthfully.

The incidents and facts mentioned have been checked but if there are any mis-statements, I shall be obliged if these are brought to my notice for clarification/amendment. I fully accept the blame and brickbats, for omissions or mis-statements and request my readers' indulgence.

For assistance in collection of information, I owe my thanks to the librarians and staff of the IDSA, USI, and Delhi Gymkhana Club.

My special thanks are due to K. N. Singh who edited the book; Tindi Sharma, N.B. Singh, Sherry Desai for correcting my drafts; and Tanwar for help with the production of maps and sketches. The book would have never ever got out without the hard work by my friend Sunny.

Finally my publisher, Vivek Garg, for his patience whilst the manuscript was amended and retyped.

Major General Partap Narain I.A. (Retd.)
M.A. (Cantab), F.I.E.

Preface

This book was inspired by a book of poems gifted by my friend's wife, the late Allan B. Moray Williams. At the end of his "Ballad of the Hundred Soldiers" on page one, he asks those of us left behind from Hitler's War—"Comrades dear comrades, what have you done so far to attain Utopia?"

We have little to show of our achievements; fun went out of soldiering half a century ago! We were pushed around from one campaign to another, made to fight with antiquated weapons as provided by the Babu–Neta Raj whilst we in military uniforms were presumed to lack judgement.

Further research into the problems of weapon production in India, led us to the conclusion that due to the lure of enormous commissions by the foreign armament manufacturers, neither the users nor purchasers are interested in indigenous weaponry. The arms manufactured by us are too expensive. Industrial revolution has bypassed India – as it has not taken advantage of the civil industrial power. We award prizes of Rs 50,000/-which interest small entrepreneurs. The in-house direction of R&D, by scientists working from 9 to 5 pm has failed to invent, or redevelop Tippu's, self-motive rockets, which caused fires. The future lies with the Universities (see Appx III).

There are many a report forecasting a bright future for India, this book brings out the danger of high value imports for Defence.

For example, whereas the Defence Ministry's report for the year lauded the HAL for test flying the intermediate Jet Trainer, and showed a twin engine MIG on its frontispiece. It suddenly changed track and ordered single engine *Hawks,* from the UK at a cost of 8000 crores—planes, to start arriving after three years.

The Times of India of December 24, 2003, reports that India is to purchase a large number of 155mm artillery guns from South Africa, Israel or Sweden. Hopefully it's wrong.

My book *Indian Arms Bazar,* "page 80." A project for manufacture of 75 complete guns and 125 spare barrels per year, on a single shift basis at Ordnance Factory, Kanpur (OFK) was put up at a cost of 81.9 lakhs in late 1959, by Admiral Shanker. His recent "*Memoirs and Memories*"on page136, confirms that a heavy gun factory at a cost of over 44 crores was built at OFK—the dedicated factory for manufacture of Artillery guns which inherited its know how from the factories in the Bengal area. It has produced 155mm gun barrels –and has a longer tradition compared to either the factories in Israel or Singapore, which are recent entrants.

Our high cost of production, helps the import lobby to say that it is "Cheaper to import than manufacture indigenously." Cost controls; are foreign to our system—where Ordnance Factories work on the basis of costs plus ten percent.

Our expertise in extracting commissions is so high that none, not one has been punished!

The system has led to import of Hawk trainers for the Airforce. Manufacture of Arjun Battle tanks—is a joke—see Appendix-1.

Our antique logistic system also offers economies without effecting the efficiency of the fighting Services. Acceptance of Change is the first requirement if we wish to usher a bright future. As a riposte to "*Tehelka* "our Government proposed to reduce costs by granting licenses to industry to supply weapons to the Armed Services. The Finance Minister, an ex-soldier, recently told the Rajya Sabha his tight financial position. Reforms could hopefully give us such superiority in military power that no evil eye shall challenge us — and we would attain peace.

About the Book

It starts with the Great War – runs through the Second World War and other wars to Buddha Smiles ushering the Atom bomb—through to Kargil and finally *Operation Parakaram,* an exercise in coercive diplomacy. It ends with *Chapter-16* and it is about how India could arm its defence forces adequately without hurting its development.

The incidents in this book are as witnessed and written by my British and Indian friends, including shorter ones by junior ranks–as a *Havildar*[1] , describe the scene as viewed closer to the battle.

The Great War 1914-18—literally consumed men. The Indian Corps lost 34,252 men in the first 13 months. Jick Rudra, Kish Norrowji who survived—the whistled orders for "over the top." Grenades, machineguns mortar and, Bangalore Torpedos of that war survive today. The need to watch the enemy is omnipresent. In 1915, the British had to barter 32,000 German binoculars for rubber.

In the year 1917, the British Parliament in recognition of the bravery of our troops agreed to delegate to Indians greater power of administration. The tank battle at Cambrai succeeded.A Tank and an airplane were sent to Lahore, which I saw.

At the end of war, Indians including Rudra and Carriappa were commissioned from Mayo College. Iskander Mirza was our first officer, out of Sandhurst.

Lord Kitchener reorganised the Indian Army. Half of it was committed to garrison the N.W Frontier, as described in "*On the Durand Line*": it provided live training ground.

Consequently our soldiers, the term includes sailors and airmen excelled and took a major part in World War II. We won 28 of 52 total Victoria Crosses, awarded. The Allied Armies were supported by American production of guns, tanks, ships and aircraft. It is said Stalin raised a toast to Roosevelt "Detroit is winning the war." America's "government owned and company operated" factories reached fantastically high rates of production—Liberator bombers at one per hour, and Liberty ships in a fortnight; in fact the record was four days 15 hours and 29 minutes.

The Germans defeated Poland and France by their speed of attack by tanks and fighters—Blitz-Krieg. They in turn were beaten by the flood of Sherman tanks, fighters, and guns from America. Flame throwers helped our two V.C's to push the Germans across the river Po and achieve victory in Italy.

In Burma shortage of British troops led to entirely Indian Brigades. Out of 27 VC's 18 were won by Indians. At the Irrawaddy, some troops of 2 British Division, refused to disembark. Yet surprisingly, at the time of celebrations of the golden anniversary of the XIV Army, our Government ruled that it was not our war.

On the staff of F.M.Slim, I learnt the importance of logistics. Operations depended on availability of *Dakotas* Jeeps, the Indian produced PBS[2] , and extra lift by boats. Our factories supplied 12 million rounds of small arms ammunition per month. LMG's, shells and fuses, guns, mortars and air bombs. India was the sole manufacturer of mountain guns in the Empire; we sent rifles to Britain and shells to Russia, by road through Meshed. Hindustan Aircraft Factory (HAL) produced drop tanks and overhauled every type of plane. (Satish Dhawan, who later headed ISRO, was an engineer with HAL).

Kashmir 1947-48

We won all the battles but lost the war despite the valour of five *Param Vir Chakra* winners. General Thimayya and Aviator Baba Mehar Singh as Nehru, did not understand, that "Big nations do not fight small wars."

Congo 1961-62

Two Infantry Brigade groups fought with exemplary conduct for the U.N. The first Brigadier Raja used Anti-tank recoilless rifle, to destroy a Dozer clad as tank. Captain Salaria of the 3/1st Gurkhas, won a PVC—leading a *Khukri* Charge.

The second Brigade under Reggie Narohna, called in the Air force. Later advanced into the mineral rich Katanga province. The battle winner was the General's *"Swagger Stick."*

The 1962 China War

Fought on two fronts by two General, my classmates, Daulat Singh in Ladakh offered credible defence. Bijji Kaul, a political appointee in NEFA, though bubbling with activity was struck down by pulmonary disease; was allowed to run the battle from his bed at Delhi! We faced disaster on several counts—wrong General, lack of a plan, non-availability of automatic rifles, poor logistics. *Indian Army's veteran 4 Division withdrew.* Defence Minister, Army Chief and Kaul resigned.

1965 Pakistan CQMH

Abdul Hamid firing a 120mm Recoilless jeep, mounted anti-tank gun, killed the leading tank destroyed two others, till the fourth tank killed him. Missile firing Saber's were downed by our indigenously produced Gnats. The Centurians of the 3rd Cavalry firing the 20 pdr guns made a graveyard of 94 Pattons.

On loss of territory in 1962 & 65 wars, the Parliament appointed the Reforms Commission to study relations between the Services and the Ministry. The report on development and production of weapons, based on three years study, has disappeared.

One of its recommendation was that Secretaries to the government should not be appointed Chairman of PSUs.

Two other recommendations, appointment of Chief of Staff and setting up of Defence Industrial College still await implementation. Defence production lapsed, we became the largest importers of weapons.

The 1971 Bangladesh War—Fought on two fronts East Pak & West

The successful crossing of River Meghna was a credit to Generals Sagat Singh and Gonsalves; they used heliborne troops and PT 76, tanks of 63 cavalry to swim across the 3-km wide Meghna River.

The Western Theatre witnessed a tank battle at which two Param Vir Chakras were won. Second Lieut Arun Kheterpal won it on a Centurion—but the queen of the battle were the T-54 trawler tanks that cleared the lanes in the minefields. Major Hoshiar Singh of the Grenadiers won the PVC for exemplary leadership in repulsing enemy counter attack.

Indian Peace Keeping Force (IPKF)

IPKF in Sri Lanka—was an unplanned, knee jerk decision by Rajiv Gandhi. I asked him why he didn't have "Think tanks"—he evaded reply. I quote Sreedhar[2], he increased Defence expenditure, pegged it at 4% of the GNP. The 80's were notorious for import of conventional arms. It started our economic downfall, which hasn't stopped. The value of the rupee slid from 9.628 to the dollar in 1982-83 within 10 years to Rs 22.7 in 1991 and to Rs 46.65 in 2004.

Kargil

It need not have been fought. We failed to replace the electronic intelligence gathering system and lack of vigilance due to bureaucratic delay. It was won by our brave infantrymen with firepower of our Artillery—in which the Bofors 155 mm and 160 mm mortars played a significant part. The Air Force entered the fray with unsuitable planes—but made up with massive bombing to keep the enemy down. We again imported heavily. The value of rupee went down from 43.1 to 44.9 a dollar within 1 year.

Buddha Smiles is about Atomic tests and **Parakaram**, our failed attempt at coercive diplomacy. Introspection into the causes of our failures, leads me to suggest that it is basically, due to the lack of education of military matters-see Appendix-III.

Epilogue presents a view of the mountainous border-future strategic defence problem. A change in the present expensive procurement system is imperative. Government's resolve to use private industry to meet defence requirements; and use of Universities is essential.

An effective way would be to hand over some Ordnance factories to the industry on basis of "Government owned, Company operated factories" following the American system. It would eliminate a lot of problems.

A wise list of what can be done right now is as under:

1. Rations in kind–The Second Pay Commission – instituted the system of supply of rations in kind to the officers. It has resulted in a miracle of achievement. We have the ability—to supply fresh a leg of chicken, 4 large potatoes, three onions and condiments to cook the same – every day – to each and every of over 40,000 officers of three services in peace stations. The system is too costly. Four years ago, the Chinese replaced it with a cash allowance.
2. Ordnance Clothing – We continue to run a 40-year-old system—Hundreds nay thousands of *Durzees* (tailors). Great grandsons – of those employed during the Great War of 1914-18 toil for long hours, get production bonus—to stitch uniforms for the Service Personnel. Working on antique cost plus ten percent profit basis they churn out uniforms at atrocious costs. A *havildar* told me the other day that he uses the regimental tailor – as they issue clothing even the colour of the Jacket does not match the colour of the trouser and alterations are necessary.
3. Ammunition Dumps—full of time—expired stocks, report fires at such depots—It is time we surveyed the situation.A number of filling and holding factories were established in constituencies. Some are so remote that their holdings would take too long to reach the battle front. If possible, all mines are filled and stored at say Khamaria – Commanders make direct indents on the factory cum stock holder. This will eliminate a number of forward Ordnance Depots. More importantly it save lives in case of a repeat of *Operation Parakaram.*
4. R&D Laboratories—The last count was 52. We keep on adding. It is time we understood that best atmosphere for advancement of technology—is at our IIT's and a few Universities and these have to be funded by Industry and Defence. The users, Industry and Professors form a triumvirate. The Intellectual property – the Patents—are

held by the Universities who have the funds to pay for taking out patents. The system followed by Oxford University could be followed.

Notes

1. Sergeant.
2. Prefabricated Bituminised Hessian.
3. Defence Expenditure in Rajiv Gandhi Years—*Strategic Analysis*, June 1980

Contents

❮1❯

The Great War 1914-18

There are hundreds of books on the "Great War" in which millions of our Jawans fought but not one by an Indian author. It, is, therefore, not surprising that not many of us know that it was the gallantry of these valiant men that paved the path to our Independence. This story is about the exploits of our Jawans. The war at sea, the sinking of ships by German "U" boats and the saga of Air Warfare are only mentioned in brief.

For helping the British in their hour of need their Parliament resolved in 1917 that "The goal of the Government was the grant of responsible Government to the people of India." As a corollary to the statement they agreed to allot ten vacancies for Indian cadets at Royal Military College at Sandhurst.

The spark for this war was the brutal murder of the Archduke of Austria, an incident far removed from India. When the Germans attacked France through Belgium on the August 3, 1914, Britain declared war against Germany. An expeditionary force of four Infantry and a Cavalry Division was sent across the Channel. They fought valiantly but suffered a great number of casualties; the Battalions were reduced to half their strength.

The British volunteer army was small; its II Corps in France was in dire need of relief. Lord Curzon overruled the Boer War restriction on using Indians to fight the "Whites." Two Infantry and two Cavalry Divisions were rushed without even waiting for woollen uniforms; a number of Animal Transport companies followed.

The Lahore Division embarked at Karachi on August 24, and arrived at Marseilles, on September 26, leaving its Sirhind Brigade at Cairo to relieve British troops guarding Suez. This too joined the battle in France within the month.

It was indeed a "Great War," as for the first time the industrialized nations moved huge conscript armies by rail. France used, 7,000 trains to move 3,781,000 troops to their posts with thousands of guns that fired millions of shells.

On October 21, the first two battalions—the 129th Baluchi's and the 57th FFR—went into battle to plug the hole in the defences held by the dismounted British Cavalry. The need for additional troops was so urgent that even 21 and 22 Field Companies of the Bengal Sappers were pressed into service in Infantry role.

Fighting in unfamiliar environments, with new rifles, wearing steel helmets, our men saw airplanes for the first time. They held the muddy trenches half full of water, in cold and drizzling, weather and fought like tigers not for their 15 rupees per month salary, but the pride of their profession —"Izzat". They gave such a good account of themselves that the Germans were afraid of fighting the "Brown Devils."

The Baluchi's, my brother in law's battalion, won two *Victoria Crosses* (V.Cs), the highest award for bravery. On October 31 they were involved in confused fighting. Two machine gun posts of the regiment were cut off. Sepoy Khudadad Khan handling one of these was left for dead along with the rest of his team. He crawled back through the barbed wire and rejoined. He was awarded the first V.C. Naik[1] Jehan Dad had a very interesting experience—he was caught in the full glare of a searchlight. With great presence of mind he walked straight into the German lines—"Slamming" all the way. The Germans expected him to be a deserter. They fed him, and asked him to get back with more of his type; but he remained stern and unfluttered . He was promoted to Havildar,[2] but was killed within the year.

Further down the line the 58th Sikhs and the Sappers of the 20 and 21 Field Companies were involved in the attack at

Neuve Chapelle in October. The Germans bolted. We took some prisoners. It is said that one of them cried—he was sure that the Sikhs would bayonet him—till a Sardar told him "Be not afraid!" in chaste Punjabi.

A large number of our braves like Subedar Mulla Singh and Havildar Karam Singh of the 58th won the Indian Order of Merit (I.O.M), which entitled them to an additional pension for life.

Naik Durwan Singh Negi of the Garhwal Rifles won the second V.C. of this War. He led the attack into the trenches from a flank—and although twice wounded in the head and the arm, he continued fighting courageously; his Company Commander saw him streaming with blood from head to foot. Lord Roberts, over 70 years of age who had raised the Garhwal Regiment, came to congratulate the battalion, dressed in full regalia of a Field Marshal. He was unfortunately killed when his horse slipped on the frozen snow and sleet.

A posthumous V.C. was won by Rifleman Gabar Singh Negi, who on death of his NCO, took command of his platoon, and ran from trench to trench and bombed each bay. He was always the first with his bayonet.

Rana Jodha of the Tehri Garhwal Sappers, who had brought his company to France to become part of the 2nd Garhwal Rifles, commanded a double Company of the battalion. During the battle of Loos in September, he was wounded but went back to the front. When again wounded by shellfire, he was evacuated and awarded an M.C. He was the first Indian King's Commissioned combatant officer. A nephew of the powerful Prime Minister of Nepal, he was appointed A.D.C to the Governor General.

During a Gas attack in April 1915, at the second battle of Ypres, Jemadar Mir Dast of the 57th—when the officers were killed—rallied his men to hold the position. For his great courage he was awarded a V.C.

The casualties of the Indian Corps in the first 13 months were 1,525 officers and 32,727 other ranks. In the bloody battle of

Flanders, the 6 JAT (1st Royal) suffered 80% casualties, the highest suffered by any battalion. Their trenches had been inundated by the Germans but in biting cold they stood in them and fought.

After the battle of Festubert

The message of commendation from the Divisional Commander was-" The earth has shaken with the tread of the "JAT." The officers except for a few Indian doctors were all British. The I.M.S. set a high standard. Both Major Atal of the Baluchi's and Capt. Inder Singh M.C of the 57th Sikh were killed. Capt. Som Dutt RMO of the Probyn's Horse, whom I knew, won an M.C. for attending the wounded under enemy fire. He also had an Iron Cross pinned on to him by a grateful German treated by him.

The two Indian Cavalry—the Meerut and Mhow—Divisions in France had little opportunity for fighting in the mounted role, they were trained in grenade throwing. Some regiments were moved up for the battle of Somme, in 1916—but saw little action.

One brave Daffadar Govind Singh of the 2nd Lancers volunteered to take a message across two miles of enemy territory. His horse was killed, but he walked and crawled across to execute his mission. He again volunteered to take back the reply. This time his horse was cut into half, behind his saddle by a shell. Yet, he again made it and lived to wear a V.C.

In view of the onset of the second winter and the subsequent hindrances of supplying rations to mixed formations, Indian Infantry Division was transferred to Middle East. King George VI paid a visit to the Indian troops to compliment them for their exemplary devotion to duty.

Tanks were first introduced into battle in September 1916. Their advent led to slow extinction of horse cavalry. The supply of fodder was a great drain on the shipping resources.

Turkey and the Middle East

The entry of Turkey into the war was almost a comic affair. The Turks had paid by public subscription for construction of the two warships at British yards. Churchill commandeered these. The incensed Turks retaliated by seizing two British ships that had sought refuge at a Turkish port. Manned by Turkish sailors and assisted by Germans in Red Fez caps, they raided the Russian port of Sebastapol. Following a formal declaration of the war, a *Jehad* was declared on the 5th November. In retaliation, Britain took over Egypt. The general offensive was against the Turks who were supported by the Germans and Austrians in the Middle East and the war in the colonies was mainly fought by Indian troops.

The Kaiser granted an interview to legendary Hardyal of the Ghadar Party. The Germans sent a mission to India via Kabul to disrupt the Indian war effort. It was stopped at Kabul, but Raja Mahendra Pratap of Mudsan, its leader, stayed on, and re-emerged at the time of the 3rd Afghan war. I met him later as an M.P. It is a great pity, that we have little information on the role of the nationalists, coinciding with the uprising in 1919—but that is another story.

Gallipoli was a disaster. Britain as a Naval power wanted to control the Bosphorus—it would also provide a warm sea route for supplies to Russia. The purely Naval attempts along with the French having failed a military operation was mounted from Egypt in April 1915, with troops from Australia and New Zealand. The Anzacs suffered enormous casualties. An Indian Brigade with Muslim troops had lesser casualties, as afraid of desertion by the Muslims in face of the "*Fatwa*," to support Muslim Turks they were not committed. Total casualties for the Empire were—213,980. Diseases too took their toll. The troops were evacuated in January 1916.

The Suez canal, the "Jugular Vein of the Empire", was divided into three sectors. It was defended by 11 and 12 Indian Divisions made up of Brigades of four battalions—one regular and three state forces. They held positions in trenches with the canal as an obstacle.

Middle East—Over 70,000 Indian troops fought in this theatre, which was under Turkish suzerainty, except Egypt. The British occupied Egypt by an unwritten convention. Lord Kitchner, the British Counsel General, who earned the title of Earl of Khartoom was like those of the other countries, but had 6,000 British troops. He wielded power with his personal prestige. The Egyptian Army was in Sudan under the Governor General of Anglo Egyptian Sudan and Sirdar of the Egyptian Army. The Sirhind Brigade, as already mentioned, replaced the British troops.

Mesopotamia was an important source of oil for the Royal Navy. To safeguard the supplies of oil, a Brigade of the Poona 6th Division was sent up the *Shatt-al-Arab* on September 29, 1914. It took Basra, and with the 12th Indian Division proceeded up the country past the famous Cestiphon. This campaign resulted in the greatest disaster to Indian Arms at Kut, some 120 miles up-stream of Basra, where some 9,000 men of the garrison were besieged by the Turks in the wet lands.

My first personal memory of the Great War is of a "Sayapa," a "Lamentation ritual" of our neighbour, a postal clerk, who perished at Kut. There were so many dying that some women had turned into professional mourners, they would make loud cries, beat their breast lead the ritual way of socially announcing the death of their client.

General Aylmer, at the head of 7 Division, arrived at Basra. He attacked on the night of January 12/13, to relieve the garrison of Kut. Our men fought with the proverbial bravery; three soldiers— Lala, Chatta Singh and Khan—won their VC's but victory eluded us. After losing 1,113 killed and wounded he withdrew, tried again on January 18-19th but lost twice as many, tried again in March with no success. The War Office suspended him. Gorringe took over; and made a last attempt on April 22. He sent up steamer "Julnar" with 270 tons of supplies. The Turks fouled its anchor. On April 29, the garrison capitulated. Whilst Townsend and the officers were sent to Turkey in steamers, the men, both British and Indian, marched across the desert, under terrible conditions—few survived to say that they were once at Kut–It finally was retaken in March 1916.

Known as the War of inventions, it spawned chemical and tank warfare, grenades, trench mortars, wireless, sound ranging, aerial ropeways, smokeless powders and antiseptics. German "Zeppelins," cigar-shaped, large, lighter than air balloons, bombed London and brought the war into British homes, till a Naval pilot—who lived to wear a V.C. shot one down. 'Madras Sappers' contribution was the "Bangalore Torpedo," for blasting a passage through barbed wire entanglements. This device was shown in the film "The Longest Day" and is still in use.

The infamous year 1916 saw the 300 day long battle of Verdun, which resulted in 3,00,000 casualties. A cemetery in France alone has 1,30,000 graves. General Sir Douglas Haig, a cavalryman brought up in the cult of offensive, was held responsible for excessive casualties. The troops had orders to get out of their trenches, and advance at the blowing of a whistle, despite murderous machine-gun fire. The U.S. Army estimate of casualties was 37.5 million, of whom 8.5 million died.

Little known is the fact that Mahatma Gandhi, as an Honorary Lieut. Colonel of the Indian Volunteer Ambulance Corps, his wife Kasturba, and Sarojini Naidu, looked after our wounded at the Netly Hospital at London. The Royal Pavilion at Brighton was converted into a nursing home for the convalescence of Indian soldiers. The funeral pyres of thousands of our braves glowed on the downs outside Brighton—where today a "Chattri" stands to remind us of their supreme sacrifice.

Allenby's Campaign

The two Indian Cavalry divisions were withdrawn to the Middle East, which witnessed the victories, of our troops, under Allenby against the Turks. When America supplied 12-inch pipes to extend the sweet water supply to Gaza, General Allenby called Allah-Nabi by the Arabs, took up operations for the capture of Gaza. He had most of the famous regiments of the Indian army to wage the only successful campaign of the year 1917.

Gaza was taken over in October. A Risaldar of the Hodson's Horse and his orderly captured 300 Turks. Lieutenant Pat Rees of the Rajputana Rifles with only six men, captured three guns. Major Dalpat Singh won an M.C, when he galloped on an enemy machine gun, killed the crew and captured the Turkish Commander.

Jerusalem was taken on December 9, 1917. Later, Haifa was captured in a daring attack by the Mysore Lancers and Jodhpur Risala led by Dalpat Singh, (who was unfortunately killed). It opened the door to Damascus. The 5th Indian Cavalry Div pursued the defeated Turks, who sued for an armistice on October 31,1918.

Most amusing was the claim of General Townsend—of Kut fame—that he arranged the armistice with the Turks. He wanted to be appointed as the British Ambassador to Turkey!

The year 1917 was a difficult year for the British. Lloyd George had taken over as the Prime Minister. The Russian Revolution broke out on March 12, when members of the Russian Duma, along with workers' leaders, set up a provisional Government. The Czar resigned. With the collapse of the Russian allies, the Germans started withdrawing their troops from the Eastern Front. They trained them as Commandos for taking part in proposed offensive, called "Michael," planned for 1918 Spring.

There was an urgent requirement of reinforcements. America had entered the war but it took time for fresh troops to come into battle. The news of the enormous casualties had slowed down the recruitment, particularly of Jat Sikhs. Mazbi Sikhs were enrolled. Besides support for the British cause started flagging. To give a fillip to recruitment, the British arranged to send a Tank and a couple of Aircrafts for demonstration to India.

The Airplanes, the first to arrive, were heralded amongst tremendous publicity. I remember how we went to see it at our Landau. The whole city had flocked to the open *'maidan'* at Mianmir cantonment. The flying display left an indelible experience. I saw the pilot in his little Biplane fly past—he waved and flew away. It was all over in minutes—in a cloud of dust.

The trapezoidal Tank with its track on the periphery was mounted as a static display in a corner of the Parade Ground—near the Lahore Fort. We, as a school, used the parade ground for games and thus saw it quite a few times.

The exhibition of new weapons was a confidence building measure, but the people were not convinced—the Germans still had an edge. In support of my statement, I can do no better than repeat my father's remarks. He had been to meet his friend Sir Mohammed Shaffi, a "*Rais*"(landlord), who lived near us. During a discussion on the outcome of the war, he told my father, "I am waiting with garlands—shall put them round the neck of whosoever is the Winner!"

The couplet by our famous poet Sir Mohammed Iqbal

"Fateh English ki hoti hai
Kadam German ka badta hai."
"It is said that the British are winning
But the Germans are advancing."

The ambivalent state of our minds was brought to fore as war clouds gathered. The British through their informers had realised that they had to do more for winning the hearts of the Indians. They realised that it was not possible to hold in perpetuity 300,000,000 Indians, with some 40,000 British soldiers. There had been mutinies of the cavalry at Lahore, followed by one during Christmas at Singapore and the affair of the "Kota Maru."

With the back to the wall, in their anxiety to win a desperate war against Germany and to improve the supply of recruits, as a thanks giving, and even possibly to appease the Americans, the British decided to take a long-term risk for short-term gain. In August, Montague, the Secretary of State for India, made his famous statement in the Parliament promising India greater autonomy. He visited India in September. Minto-Morley Reforms and the opening of the officer ranks to Indians followed.

It was a sea change, considering that in 1914, Jick Rudra, when he went to join the Army in London, was told that he was not eligible. It was only when he returned without "*pugree*" that the sergeant took him in as he held a Ceylonese passport.

H.S. Malik, a student at Oxford, was not permitted to join the Royal Flying Corps. His tutor sent him as an ambulance driver to France and when the French offered to enroll him, that Mr. Urquat, his tutor, wrote to General Henderson and Malik was commissioned. He shot down several German planes, and was wounded during his illustrious flying career. Many others followed him into the officer ranks.

Interestingly, Lloyd George also took another historical decision–he announced the Balfour Declaration of 1917. This gesture to the powerful Jewish lobby in the U.S.A—helped in obtaining funds by donations. He opened the gates for immigration into Israel. The number of Jews in Palestine went up from 85,000 to 600,000. The gesture could also be thanks giving to Dr Wiezman for his invention, on production of Acetone!

For the dreaded attack "Michael" in March 1918, the Germans used eighty Divisions, supported by Poison Gas, 5,473 guns and 3,535 mortars. They used "Big Berthas" guns with barrels longer than a ten-storey high building. They lobbed a shell 4 km high, which took 176 seconds to reach Paris, 125 kms away and inflicted 38,512 casualties on the Allies.

To stem the tide American troops poured into Europe at the rate of 1,20,000 per month. An American Division though not so well trained, had 17,000 men as compared to 7,000 of a German Division.

A further German offensive with an effeminate name of "Georgette" was planned as a second act. To counter it, the Allies mounted an offensive after July 14 parade at Paris. The French learnt of the timing of attack, from German prisoners. So cleverly the allied attack was timed to coincide with "Georgette." The German guns were targeted even before they could open up. The attack was led by French Moroccan and Senegalese troops with the US First and Second Divisions on the flanks. It went through and married with the U.S. Marines in Tanks. This was the last laugh by Ludendorf—as 10,000 Germans were taken as prisoners.

The war ended at 1100 hrs on the 11th day of the November month of the year. At the end of the Great War when Jick Rudra and Kish Naurowjee, grandson of Dadabhai Naurowjee, who had survived the war, went before Montague, the Secretary of State for India, who stood up to thank them for fighting for his country and sent them to Daly College, Indore, for commissioning: despite objections of the War Office. There, they met several others, including Carriappa (Later Field Marshal) who were commissioned into the Indian Army.

Life of an Indian officer was, however, not easy. Kish Naurowjee resigned. Jick Rudra on joining the 4/15th (old 28th) Punjabis, at Golan Heights in Palestine, was allotted a small tent, where no one spoke to him. Such was the prejudice against Indian officer by the British. He sat all day, went to the mess to eat, and slept. Till the third day, when the Subedar Major—Mohd Nawaz Khan—came to his tent and told him not to worry.

"Jahan aap ka pasina girega
Hum Khun daingay."

"Where your sweat drops we will shed our blood."

Such noble words, were justified by the bravery of our men—the daring deeds of men like H.S. Malik, Roy and others who volunteered for service in the Royal Flying Corps. India's contribution in terms of manpower, was 1,457,000 men including non-combatants, who went overseas and fought alongside people of many races with diverse backgrounds. The confidence of our brave soldiers who took on and defeated the world's best soldiers on their home ground, earned India laurels and put us on the road to Independence.

Out of the gates of the Royal Military College Sandhurst came out our first officer—none other than Iskander Mirza—later the President of Pakistan.

Other Theatres

Our Jawans fought in several places as far apart as the German East Africa, south of Mombassa below the Equator on the Indian Ocean, to Merv in Russia.

This campaign in German East Africa, later known as Tanganika and then Tanzania, is of interest as our large "B Force" of some 60,000 armed with bolt operated rifles was held by some 4,250 Askaris armed with German Maxim guns.

This story is built around the personality of the German Commander of this colony. A Lt. Col. Von Lettow-Vorbeck—a veteran with 25 years service in the colony and aged 44 in 1914. He had served alongside the British in Peking during the

Boxer Rebellion (Opium War) in China and learnt guerilla warfare under the able guidance of General Bootha in the Boer War. With a small force of only fourteen companies—a total of 216 German officers and 4,250 Askaris, he kept the British Indian forces at bay for four years. Our troops under the command of Maj General Aitken commanding Force B—included the Kashmir State Gurkhas, the Loyal North Lancashire, 63rd Palmacottah Light Infantry, 13 Rajput, Bombay Grenadiers and 61 Pioneers.

They had just landed at the hot and steamy beach at Raskone near the town of Tanga, a small port and railhead. Having been cooped up on board their transports for 2,500 miles across the Indian Ocean—they were not in the best of condition—nor were our crack troops. They were largely Imperial Services soldiers from the Indian States.

When the information was received about a British Indian force of some 60,000, from Mombasa on November 1, 1914, with the goal of taking over the port of Tanga, from where they were to advance up the railway to Moshi, at the foot of the famous Kilamanjaro Mountain—the snow covered peak at the Equator.

There was only one company of Askaris at Tanga when two battalions of our troops landed. The German Colonel assembled all his troops from Moshi and moved them by rail to combat the fierce foe.

Our troops were attacked by the Germans led Askaris—equipped with Maxims. The Askaris just down from the hills, were fresh and ably led by the Germans with blackened faces. The abrupt dit-a dit-dit of the Maxims was disastrous for the Palam Kotta Infantry, which was moving towards the centre of 6th Company's sector. They turned and took to flight to their landing beach, throwing down their weapons. The Askaris with their shouts of "Wahindi-ni-wadudu" were jubilant.

The position on the outskirts of the town was however different. The Gurkhas of the Kashmir Rifles had already sliced quite a few enemies with their Khukris. They had also got to the Hotel Kaiser and hoisted their flag after hauling down the German flag. On their flank the Lancashires had broken through and managed to link with the Gurkhas.

A counter attack by two fresh companies into the centre of the town with blazing machine-guns and the springing into action by the German volunteers restored the centre. The 101 Bombay Grenadiers moved into the gap left by the Palam Kotta Infantry. Lettow ordered his number 13 Company to rush in with four MGs on their left flank and thus reduced the battalion to a company in no time. The Grenadiers, brave as they were, fought back for nothing.

While the 13 Company was straining to keep up with its counter attack, the gods of war, smiled on Lettow. The local custom is to hang beehives on tree limbs. A number of these hives had been struck with stray bullets. The infuriated bees buzzed forth to launch an attack of their own on the unfortunate troops. The African bees are like leopards to a tabby cat compared to European bees. In no time there was a general retreat to the beaches.

Next morning, troops returned to Mombasa to fight another day. They left behind sixteen machine-guns, enough rifles to equip three German companies, 600,000 rounds of ammunition, field telephones and large stocks of blankets and clothing.

Maj General Aitkin also lost about 800 dead and 500 wounded the loss of morale suffered in this engagement—their belief in spooks—enabled Von Lettow to carry on a guerilla war, for the next four years.

Two of our regular battalions, the 129 Balauchis and 57 Frontier Force Rifles were inducted on their relief from France to fight in Tanganika, thousands of miles away, south of the Equator where they served till the end of a different kind of war. The climate varied from hot and steamy beaches to views of snow covered peak of famous Kilamanjadaro mountain at the Equator.

A new type of foe was encountered, the infuriated bees sallying forth from their beehives in the jungle. Von Lettow finally signed a Peace Treaty on November 25, 1918—(a fortnight later) only after it had become obligatory for him to conform to the General Armistice.

Merv

As a consequence of the Russian Revolution, their assistance to the Turks in Caucasia collapsed. The small British force that had intervened withdrew to Krasnovodsk on the Caspian Sea. From there a single railway line runs east, via Askabad to Tashkent.

Our troops operated in the Trans Caspian region from Persia to guard this invasion route to India. They operated out of Meshed towards the single railway line from the Caspian Sea, via Askabad to Tashkent. The Punjabis, supported by Cavalry, captured Merv on November 1. The senior Cavalry Regiment—the 7 Light Cavalry—cherished its battle honour 'Merv' for fighting the White Russians.

Ruling Princes of Indian States

Indian States, forces and ruling princes took an active part during the war. The longest in France were the Maharaja of Bikaner, Sir Ganga Singh and Sir Partap Singh of Jodhpur, the favourite of the British Raj—since the raising of the Imperial Cadet Corps by Lord Curzon. Their contribution was highly appreciated as it showed the loyalty of the Indians towards their sovereign. As many as 18 ruling princes took to the field during the war. How much these Chiefs helped in running the operations is open to question— one of the smaller rulers delivered messages emanating from Corps HQ. Their exploits were faithfully recorded in book called " *India's Services in the War.*" It even recorded donations by states as little as Rs. 100 to the Y.M.C.A by the Raja of Tippera—175th in the dynasty—he had a military force of 250 and police of only 341.

A photograph of the Maharaja of Bikaner, Sir Ganga Singh standing in his uniform of a General, at the venue of the Peace Conference at Versailles in France showed that at last we, the Indians, had arrived.

Lokmanaya Tilak's call "Swaraj is my birth right" motivated and moved our people for independence which was attained under the guidance of the great saint soldier[3] Mahatma Gandhi—within thirty years of the end of the Great War.

Notes

1. Corporal
2. Sergeant
3. Gandhiji served as a stretcher-bearer in the Boer War 1900.

❮2❯

On the Durand Line—70 Years Ago

Kumar Kochhar was one of the finest gentlemen, I knew. We joined the army together. He had been to the Royal Military College Dehra Dun and had a head start on me. His father Colonel Kochhar, a distinguished army doctor, arranged for our training with the 8 Punjab and Probyns' Horse at Lahore in drill square bashing and riding.

Of all the Indians, Chinese, Iraqis and other non-British who went to the Royal Military College, Sandhurst, or Royal Military Academy, Woolwich—(The Shop), Kumar was the only Indian cadet to be appointed an under-officer.

We got back to India after nearly five years—he went straight to join 15 Field Company—which had been selected for Indianisation, whilst I drove from Bombay to spend a few days with my parents at Lahore.

One day, the local daily Civil and Military Gazette came out with the report of the Waziristan operations against the Faqir of Ipi who was gathering his lashkars for a fight. Always envious of Kumar I couldn't face the prospects of not having taken part in this little war, depriving myself of a medal. So I wrote to the Headquarters at Bangalore and got posting orders also to join 15 Field Company at Razmak.

It was good to get away from the headhunting fond mothers, looking for suitable husbands for their daughters. I shall never forget one over-powering lady, who only wanted me to say yes;

she would then call me to tea to meet her most beautiful daughter, whose main qualification was she didn't even know how to boil water, a grim prospect for an old boy-scout!

Travelling to the frontier was an experience—from Lahore by the Frontier Mail to Rawalpindi and then by car past Kohat to Bannu, night at the mess and then by car in the tribal area which the CRE had arranged complete with two *Khassadars* as escorts.

I reported to my OC Major E.H.T. Gayer. He was our instructor at the Royal Military Academy, Woolwich a hand-picked officer for commanding the first Indianised Field Company.

The same afternoon he told me to set up a water point for the new arrivals outside the perimeter. I calculated the number of men and materials required, as per the book. I needed 20 men, but was told that the Company was fully deployed. I should ask the Subedar Sahib, Mannu-Swami, to let me have the sick and other stragglers in the lines for the job. The water troughs for the mules etc. were taken to the site to be connected up to the standpipe at the camp-site, the work-party got down to work as explained. Inspite of my precise planning, I discovered that I had forgotten the ubiquitous white-tape for traffic control. I went back to the camp to draw it. On returning half an hour later, to my horror I found the Sappers watching a football match. I was furious with the NCO—but he said that the work was finished (*Ai Pocha!*). To this day, I remember the dedication to the work of Madras Sappers.

The tribal areas including Waziristan lie between the North West Province and Afghanistan. These inhospitable and bare hills are inhabited by a number of tribes—Afridis, Mahsuds, Wazirs, Baluchis etc. ,generally called Pathans. They have been independent from times immemorial and have only been marginally ruled from Kabul. It is perhaps not well known that even Akbar the Great, whose kingdom extended from Afghanistan to River Narbda, did not exercise actual control over the territory below Kabul down to the Indus: for the first 28 years of his rule because Humayun on his death-bed in 1556, had given the kingdom of Kabul to Hakim, a three year old son from his youngest wife.

This territory which has never been subjugated is inhabited by fierce highly proud tribesmen whose main ambition is to own a rifle and move about the hills seeking a prey on the highways. Lord Curzon had visited Kabul as an MP; on becoming the Viceroy, he initiated the Forward Policy. Tribal areas administration was placed directly under the Central Government, instead of the then Governor of Punjab. After the first Indo-Afghan War in 1919-21, Razmak, Wana and other Central points in Waziristan were occupied. The control of the tribal area was exercised through the political department with political officers posted to various areas who were backed by local retainers not in uniform but on regular payroll called *Khassadars.* They were in charge of keeping peace along the roads, and providing work parties to the MES for maintenance of roads.

Next came the guides and scouts—para military forces—dressed in *Mazri* (grey) uniforms, officered by specially selected *Pushto* speaking British officers. Battalions were organised in class companies of different tribes — Mahsuds, Wazirs, Afridies and other—each under Subedars and Jemadars, from their clans. The guides and scouts were well disciplined—they had regular stations with well-guarded armouries—life was tense at times when the weapons were stolen, these had to be recovered. They were ultimately backed by the regular army.

The political officers had judicial powers and administered their areas through the *Maliks* (Headmen) and occasionally held *Jirgahs* (gathering of particular tribes) to deal with law and order problems. Justice was quick as per special tribal laws.

Razmak, cantonment for the Brigade plus, under the command of General Norton of Everest fame, is located just short of Durand Line—the recognised border between India (now Pakistan) and Afghanistan. It is a bleak and cold place at a height of 6500 feet. The accommodation was built of concrete block walls and corrugated, galvanised steel sheet roofs. The Brigade Headquarters was a double storey building on a wide tarred road. The camp had a double random stone perimeter wall, the watch towers and gates of stonework were continuously improved by the Sappers. The camp was defended day and night by picquets on the hill tops and watch towers—a strictly male domain—I believed no woman had ever stepped in this sanctum.

A *kutchha* airstrip lay on its east where our No. 1 squadron of the Indian Air Force would sometime land their Wapitis, noisy propeller driven bi-planes. The playing grounds were on the west. There was also a local Civil Headquarters, office of the A.G.E. Razmak and the transport lorries of Harichand Kapur & Sons who ran the service from Bannu to Razmak.

King Commissioned Indian Officers were new entrants to the scene. The doctors had been in the Indian Army since World War I—Behari Kapur and Puri were at Razmak with the Field Ambulance cum Hospital before the influx of the infantry officers from Sandhurst/IMA to the Indianised 6/13 Frontier Force Rifles—the famous Piffers. This was followed by Sappers led by Aserappa, and Mangat Rai. We got on very well together, the troops looked up to us, the British officers watched us like hawks. The Viceroy's Commissioned Officers took time to get over their suspicions.

Mohammed Musa, who later became Chief of the Pakistan Army, Akbar "the General Tariq," who led the Pakistan tribes attack on Kashmir in 1947, Mohammed Yusuf, whom I had known as a young scout at the world jamboree at Birkenhead in 1929, Pran Narang who unfortunately was killed in World War II, were great company. We had some excellent *Desi Khanas* in the 6/13 mess—the Sunday curry lunches were very popular with all. Another activity was bridge; we played auction bridge those days!

The daily regime was fairly hectic, we got up for early morning PT parade, ended in the evening with games-basketball or volley-ball inside the camp area, hockey and football matches on the grounds outside the perimeter. The Madras Sappers as usual did well at hockey, we won a challenge trophy. There was even a golf course with 'Browns' on the west around the airstrip.

The camp was well protected; tops of surrounding hills had picquet posts—round stone wall sangars held day and night by the men from the Infantry battalions. One dark night, the tribesmen crawled up to a post held by the 1/9 Gurkhas and managed to surprise the sentry and inflicted a gash across his stomach. His *'sathis'* awoke, took out their *khukris* and let the invaders have it. The Gurkhas, once their ire is on, and have unsheathed their *khukris*, are formidable; by tradition a *khukri* is never sheathed till blood is drawn—even from one's own finger! They cut down the heads; arms whatever came in their way. The enemy fled taking away their wounds—next morning, the amount of blood around the post had to be seen to be believed. The sentry when he was brought to the field ambulance hospital had his intestines hanging out. Behari Kapoor performed an excellent operation, sewed him up and the Jawan was back hale and hearty soon after. I believe the Gurkhas behaved in a similar exemplary way in the Falklands War in recent past which came to an abrupt end soon after they landed.

The Brigade was entirely based on animal transport. Motor transport was allocated when required. The Company equipment except for a petrol engine driven Warsop drill for boring holes for demolitions and a Petter engine for lighting was archaic. The picks and shovels were kept highly burnished under the watchful eyes of the British NCOs. We had horses and mules on our establishment—saddlery and special harness (*Khajawas*) to carry the equipment on the mules.

It was bitterly cold—we would gather in the evening, sit outside on the wall, wrapped up in Poshteens and have a 'whisky mac'—a concoction of whisky and brandy developed by Capt Lillie, our second in command, before going to the 1/9 Gurkha mess for dinner, just over the boundary wall. I was a confirmed beer drinker, I disliked whisky. When served on mess nights in the Chatham mess I used to quietly pour it into the potted palms. I would gulp the whisky mac as a medicine—looking back it was like tossing back vodka.

Sappers had besides training and education—maintenance and road improvement tasks. We once had to build a high retaining wall which involved cutting a toe foundation, Jemadar

Subbarayan was in-charge; when he was ready to start building he showed it to me – a green horn engineer, I said it ought to be deeper. Within minutes of the additional dig, the whole hillside started slipping – after that I never questioned his practical judgement.

Another interesting job was realignment of a hairpin bend—Greenwood's Corner—named after CRE. We always had half section deployed for local defence. The Sapper manpower was therefore limited; the local coolies when arranged by the department were a lazy lot, consisting of young boys and old men. Two men with one shovel between them—one holding the shovel and the other pulling it with a rope—were less than half as effective as a sapper. The infantry working parties were effective. I shall never forget the 2/11 Sikhs—the famous Royal Ludhiana Sikhs, excellent specimen of men, all six feet, well led by the British Officers who always worked alongside—they could shift the earth really fast. The first time they worked for me—on the new road from Razmak to Razani—they finished the four hour task in less than two hours—I therefore increased it—despite protests by Capt. McLoed, the O/C detachment. They got down to it with their war cry *Sat Siri Akal* and marched off well before time.

The Faqir of Ipi was a Tori Khel, with name Mirza Ali Khan; he came into prominence by championing the cause of ISLAM, starting with the case of Islam Bibi, a minor Hindu girl, who eloped from Bannu with a young Muslim student. On February 26, he made a most inflammatory speech asking the Maliks and *Khassadars* to desert government service. We went out for various route marches called 'Columns'. The system of passing Columns through ravines was the same as practiced by Pollock in 1882 when he forced the Khyber Pass. The Columns consisted of all arms, with conventional advanced guards, main force and the rear guard—except that we started with a large advanced guard and ended with a much larger rear guard. The advanced Guard Commander used to carry a double viewing stick, he would call up one of the officers and detail him to capture the *Tekri* (hilltop) indicated, the detachment would double up the hillside to hold it till the rear guard had passed through the defile safely and then join the rear guard. The pace of the Column was the regulation three mile an hour with ten minutes halt to the hour. As precaution against the snipers, officers wore the same headdress as the troops, our sappers wore cardboard *shakos*; we had to wear pith helmets and presented conspicuous targets on our horses. I shall never forget the first shot which whizzed past my head. We had to be vigilant all the time. Excellent training under conditions leading to active service; with a system of regular reliefs–the whole army benefited.

Most of the Columns mounted by Razmak Brigade were for showing the flag—the show of efficacy of our mountain guns—3.7 Pack Hows was very impressive. The guns were carried on specially selected mules; one carried two wheels, another the breach, etc. The mountain gunners, specially selected strongmen, could slap the guns together in minutes, fire a few rounds and just as quickly dismantle and swing up the parts on the mules.

The Pathans and Rajputs have a common heritage and code of conduct. They carry animosities for generations, shelter their guests whatever their status, are extremely hospitable, sometimes embarrassingly, so they insist on peeling the egg for your breakfast with their not so clean hands!

Their villages have look out towers—like the water towers in the West, sometimes two. We had the unpleasant task of destroying a village of the *Tori Khels*, who had refused to punish the culprits alleged to have shot one of our political officers. The ultimate step was taken after all other methods of persuasion and coercion including holding a *Jirgah* (meeting of the clans) failed. The villagers were warned to evacuate, to make sure, pamphlets were dropped by the Air Force. RAF Army Co-operation and Fighter Bomber Squadrons were used for bombing, strafing and air observation. The tribesmen could only shoot at low fliers with their rifles. Jumbo Mazumdar, Subroto and Awan of No. 1 Squadron based at Miranshah, only a few minutes flying, flew overhead when we reached our target.

As usual we assembled outside the camp area in the dark, and started at dawn, going through the *Sahor-Tangi* (defile), picketing the hills. The village lay in a valley—near a stream and consisted of twenty or so houses, mainly mud walls and thatched roofs. Only the house of the headman—Malik—who was richer in resources having worked with the MES, was a bit more substantial with corrugated iron-sheets for roofing. The watch tower was a square double-storey structure.

The infantry surrounded the village which was deserted, sappers went in with their gun cotton and gelignite charges and blew up the houses, this was possibly more a political show of strength—blowing up the houses with the consequent noise of explosions. Finally, the houses were set on fire and from our ridge we saw them burning, with the wind from the west rising, the tribesmen fired desultory shots at us from a long distance. With the high cost of ammunition, they don't like to waste it—these shots were a token expression of their resentment. A quick and orderly withdrawal—well executed followed, and we were back to the camp for dinner.

The ding-dong skirmishes went on for quite sometime, till Ipis men laid a siege round the camp, cut off our telephone lines and the road. There we were, struck on top of Razmak hill with the next post at Dosalli on the Khaisora river about 35 miles down the hill—with only the occasional Wapiti dropping the mail.

The General decided that he had enough and asked the sappers to reconnoitre and re-establish communications. Lillie and Kochhar were at Dosalli, Harkirat had been posted out, only Lawny Gayer, Aserappa and myself were at Razmak—I was detailed.

Starting at dawn with the usual commotion caused by mules kicking, we started down the hill. My escort, the Gurkhas the winners of the Khud races—were extremely agile running down hill, took the shortcuts, whereas I had to walk along the road full 35 miles measuring gaps—of bridges and culverts blown, noting the telephone posts destroyed. On top of it, I was dressed in battle order, breeches and boots with map cases, binoculars, compass, camera to take photographs, haversack, water bottle festooned like a Christmas tree. By noon we got down the hills and stopped for a short break for lunch, then came the long haul from the foothills to Dosalli, the last ten miles or so—the sun beat mercilessly and the breeze died down with the result that I still remember very vividly putting my hand in my haversack and finding the chocolate bar gooey and messy. The water bottle was long empty.

The sight of Dosalli on Khaisora River was good for the eyes—we were welcomed by the garrison; there was a permanent Tochi Scouts post, plenty of water, Captain Lillie and Kochhar's detachment were building Braithwaite Tank Towers. Kumar looked after me extremely well, we sat by a small swimming pool and regaled ourselves with tea.

The night, as usual came down suddenly and just as I was getting back on my feet—we got a signal from Waziristan District Headquarters calling for the report immediately. Believe it or not! I was dead beat and was in no condition to sit down and write a report. Kumar came up with the solution—have a whisky soda—as a medicine, the report was written and dispatched in time.

The sequel of the story of how important it is for the Military to have a drink to revive them at times of great stress—reached the ears of our Prime Minister—so a few years later when prohibition was slated–we the soldiers were exempted even in Morarji's Bombay, Mumbai of today.

The Shahur Tangi[1] — April 1937

By Col. WGS Lawrie

In March 1937, Wana Garrison included battalions of the 8 Punjab, 6 Rajputana Rifles and the 3 Gurkhas plus other arms. Wana was reached from Manzai, the railhead by a road about 60 miles long, through hilly but fairly open country. Roughly 20 miles from Manzai was the fortress of Jandola and the same distance from Wana was Sarwekai. Both were manned by the

South Waziristan Scouts, while Manzai was held by a battalion of the 12 Frontier Force Regiment.

About halfway between Jandola and Sarwekai was Shahur Tangi—the defile of the River Shahur . It was several miles long following the winding course of the river, and formed a series of S-bend; each cut off from its neighbours by rocky slopes. There had been no trouble along this road for several years and British families from Dera Ismail Khan had been coming here for picnics without escorts.

At the beginning of April 1937, one of the battalions of Wana Garrison was due to be relieved; the new battalion had arrived at Manzai by train and was waiting for orders to proceed to Wana by road. The Scouts at Jandola reported unusual movements of tribesmen along the road, which was strange, as there were no villages nearby.

The RAF flew some sorties along the road and also reported some movement, but it was difficult for them to be exact, since gray and *khaki* clothes of the Pathans merged into the background of rock and sand.

The District Commander ordered the convoy to be delayed for a few days, but no further news came in. Manzai was extremely hot and unpleasant and there was no room for an extra battalion, so he ordered the move to go ahead.

On April, the 14 'reveille was at 4 am. The officers had a "farewell" drink in the Mess and they marched off at 5 a.m. The battalion was embussed in some 50 closed 3-ton trucks. A troop of four armoured cars with British troops accompanied the convoy and also a company of the 12 Frontier Force Regiment, two or three trucks with each armoured car, uncovered and filled with men ready for action. So there was a fighting party at the front and back and one-third and two-third of the way down the line. There were also a few private cars and taxis with British officers returning from leave.

Troops were standing by at instant readiness at Wana, Sarwekai, Jandola and Manzai and the RAF sent up sorties at first light. The convoy chugged slowly up the hill and reached the defile at 6 am as dawn was breaking. As they moved through the

Tangi there was no sign of life. When the leading vehicle reached the end of the defile, the convoy must have been strung out over several miles with two or three vehicles in each re-entrant out of sight of the others. Coming the other way was a'*powenda*' caravan of perhaps 50 laden camels with a dozen drivers. On a given signal the drivers turned their camels loose across the road and the convoy halted in confusion. The timing was perfect.

It later transpired that a lashkar of about 4000 men had been sitting out on the hillside for over a week with no cover, bedding, or campfire or any of the impedimenta that regular troops would have required. As soon as the convoy ground to a halt, the Pathans who had been hiding in the ditches stood up and fired at point-blank range into the covered lorries where the men were sleepy and car-sick. The officers in private cars or taxis were shot through the windows.

Where there were troops and armoured cars ready for action they put up a stiff resistance, but the armoured cars were designed for the desert and could not aim their machine-guns very far upwards or downwards. In some re-entrants there were three or four covered lorries and no men ready, to fight. These suffered heavy casualties, losing. Hundreds of rifles and thousands of rounds of ammunition were looted..

This was before the days of universal wireless sets and there could be no effective control of the battle. Luckily, the leading two or three vehicles got away and dashed on to give the alarm at Sarwekai. From there they could telephone to Wana, Jandola and Manzai, but it was mid-day before any sizeable force could be deployed on the ground. The RAF was helpless, unable to distinguish friend from foe. The battle ranged till nightfall. There were many deeds of gallantry which were later recognised by awards of DSOs, MCs and IDSMs but incredible losses of men and weapons. Seven British officers were killed including several of my friends.

The Captured arms and ammunitions enabled the Pathans, incited by the *Faqir of Ipi*, a "holy" man in the pay of Hitler, to embark on a long campaign known as "Waziristan 1937-39" which dragged on for the next two years and required at its height 60,000 troops to put down. An advantage was that the Indian Army became expert in the techniques of mountain warfare, which stood them in good stead in the subsequent campaigns in Abyssinia, Tunisia and Italy.

I was very lucky to arrive in Manzai on posting to Wana the day after this affair, when the whole garrison was still shaken by the horror of it. I went on to Wana in the armoured car and we only got through by pushing still smouldering lorries down the hillsides. After this, the road was closed for the next two years and Wana had to be supplied by air.

PS: The commander of the convoy was "Chatty" Paton whom you will remember at the Shop Woolwich. He was awarded the DSO. One of the officers killed was Eric Hinde who was at the Shop with us.

Note

1. Defile

❮3❯

Penang was Fun

War can land one in strange and awful places, but Mars the God of War, was in a benign mood, I was sent to Penang. People went and still go for holidays to this *Paradise Island* also called the *Prince of Wales Island* by the British. Its real name came from Pulau Penang, the betel nut tree.

Founded by an Englishman, Francis Light — born out of wedlock—when the son of a rich farmer of Theberton, got his mother's maid Mary Light into trouble. He was packed off to the Navy as soon as he was old enough. After a few years in India, he set up as a merchant. With his persuasive powers, he negotiated its take over, from the Sultan of Kedah. Finding that his labour was greatly disheartened by the thick and tangled jungle, snakes, poisonous reptiles and fevers; he, one day, called his labour together, filled a large cannon with silver dollars and fired it directly into the trees. His Tamil labour took out their axes and fell to it—and a Colony was born.

It was in 1939, just before the start of the Second World War, the 40 Pathans and my Section of 15 Field Company Madras Sappers were detailed to hold and defend this far-flung corner of the Empire. The 'Pathans', the traditional name of the 1/16 Punjab Regiment, was a misnomer. In fact the battalion was a mixed class unit. The 'Pathans' included a company each of Sikhs, Dogras and Punjabi Mussalmans—A regular battalion, with excellent officers, and well-trained men. They looked very smart on parade in their grey Mazri bush-shirts, with black buttons.

I was lucky to start the war on Penang's ideal beaches. There are an unlimited variety of beautiful coves and beaches, which are easily approached by the circular road around the Island. I do not recollect ever having done the complete round, there were so many inviting places that one never really went too far. One snag was that the beaches harboured a stinging flat mud fish, when trod on, it's sting sent one reeling with pain. The answer was to wear plimsoles—as long as the feet were safeguarded there was no need to worry about dress or undress. The beaches were so secluded that one could have nudist picnics as long as one did not have qualms about being watched by some small time thief, who could be hiding in the jungle, to pick up a camera or a lady's handbag, while the party was gamboling out in the lovely sea.

Those days, the island was approached by the Butterworth ferry from the mainland. Prominent buildings were the Government House and the Eastern and Oriental (E&0) Hotel on the main Farqhar Street. It stood within a few yards of the sea wall. The George Town, the Runneymeade Hotel was near the Penang Club. Fairly well developed civic amenities included a civil hospital and an aerodrome. There were also the usual *bazaars.*

For entertainment there were two cinemas, showing English pictures. The Great World and the Happy World dance halls offered non-stop dancing. These were like the "Dance Palaces" of those days in England. Bands played non-stop, tables were arranged around the dance floor. Entrance was by tickets, partners were available—young Chinese or Eurasian girls, who were always well dressed, some even spoke a few words of English including "I love you." It was said that the girls took on this profession to earn enough money to buy their trousseaus before getting married. They did not date or go out with the clients. For love life, in Malay the *'Sahib log'* had keeps, who were installed in flats and were bought and sold at the time of transfers.

The tickets were cheap at 50 cents, so were the drinks. The Malay dollar was worth about one and a half Rupee. After a couple of months, dancing at these public places—there were at least two at Singapore—was discouraged, as a possible source of leakage of information.

A funicular railway ran up to Crag's Hotel—2,400 feet, the highest point on the Island. For a small ticket one went up and could breathe the cool and clean air, feel the mist. Surprisingly, the number of people going up to the peak was small, that may be due to its distance from the town.

We, the military lived on a plateau, away from the town. The main buildings were a magnificent double-storeyed Royal Artillery Officers' mess, at Glugor barracks. A Royal Artillery Regiment, once manned the guns of the fixed defences. These had since been handed over to the Malay Volunteers, British and Eurasian planters and tin miners. The mess building was well appointed, the furniture was teak and mahogany; old bearers and staff of the 40 Pathans ran it most efficiently.

Our bedrooms were on the first floor. The dining hall and anterooms including a ladies room were on the ground floor. Amenities included a billiard table and two squash courts. There were a few married bungalows, with of course the CRE's at the best location. The garrison of Penang consisted of the 40 Pathans and my detachment (64 men) of 15 Field Company of the Madras Sappers, with medical and other services.

War Declared

It was during a peaceful Sunday afternoon tea (September 3) that we heard on the crackling radio, the BBC news of declaration of war—there was hardly any comment. As the only regular engineer troops, our first job was to commission the Search Light Station. Recalling the training received at the School of Searchlights Gosport, we got the engines at Batu Maung working and handed over a working system to the local volunteers. Unfortunately, the guns were never fired in anger. The Japanese just walked in, captured Penang, and took possession of stocks of tin, rubber, the ships, tugs and all.

Penang was an indolent station; we exercised to keep fit. I shall never forget a young Mahsud boy, being hit on the back of his head by a speeding bus, while overtaking a squad of our men on a training run. He was in the last file of a group coming back to camp after a cross-country run. His head split, the brains came out. We learnt our lesson to run on the right of the road, facing the traffic.

Our medical officer Paddy O. Driscoll started protecting us against tetanus. Those days, it was in two doses. I recall, how after the first prick I broke out with skin eruptions. There was a party in the Sisters' Mess that evening. Paddy gave me yet another injection to make me fit for the party.

Singapore and Penang were then both treated as married stations. I was asked to receive my friend Captain Pran Narang's young wife and to send her on by train to Singapore. We had been together at Razmak when he was in the 6/13 Frontier Force Rifles—the Piffers 59th. Pran looked every inch a Pathan in his dove grey short silk *pugree* tied in their special twisted style—on a gold *kullah*. His matching character, an open heart he was with a devil may care expression, he was excellent company.

Alas! He was nonchalant to the extreme. He was killed in the Arakan where he went into the battle without his steel helmet. I was looking forward to meet the lady till I got a call, to accompany the Fortress Commander on a reconnaissance. The old men wanted to look into the defences of his Parish. Her steamer was due to arrive at a time when I would be away. I, therefore, arranged with my friend Gurcharan Singh Virk, a leading lawyer to receive her. After lunch, my friend brought her over to the mess and handed over charge to me.

Bado, as Mrs. Narang was known, was also an open hearted person, an extrovert. I can still picture her, decked up in a chiffon sari—pinkish—it made her look even bigger. I just managed to stop her at the broad sweeping staircase. I didn't know that she was with a child—it makes women look prettier. She told me, she had a lot of difficulty getting her family's permission to travel—in the end she said—She was going any way.

We had tea in the ladies' room. I took her to the railway station on the mainland, crossing over in the ferry. The steamer from India having arrived, the station was crowded, there was a lot of jostling and commotion. Suddenly, a beggar fell down on the platform and started throwing an epileptic fit. In the commotion, one of the lady passengers, who was watching it, found that someone had snatched her handbag and got lost in the crowd. The police was called. They immediately suspected the person having the fit, as an accomplice. He was writhing on the platform, with a couple of dollar bills in his hand. These possibly went to the police.

I was glad that my ward was installed in her sleeper for the overnight journey. Pran, as the Brigade's DQ (Deputy Quartermaster), had no doubt allotted himself a suitable bungalow.

Planning Malay Style

Accompanying the Fortress Commander was the most important duty. I got up bright and early. On reaching the Fortress Commander's office, the Brigadier very kindly offered me a glass of whisky, which I politely declined. There were two main axes of enemy attack, through Signora in the North and Kota Bharu on the South China Sea in the East. He had asked me to accompany him on a reconnaissance of the Kota Bharu axis.

We set off across the ferry to Butterworth. From there we drove right across to Kota Bharu. (Major Japanese landings took place on the beaches of Kota Bharu. The battleship *"Prince of Wales "* and cruiser *"Repulse"* were sunk nearby). It was a fairly good tarred road about 22 feet wide. After sometime, just before we reached the seaside, there was a small hill or a sand dune through which the road wound on to the small township. The Commander immediately stopped, and selected it as a defensive position. I got down to working out details of camouflet[1] charges and advised him that it should be possible to impose a delay of two hours. The Brigadier was most unhappy. He told me in no uncertain terms that he did not want this road at all. We came to an impasse, till we drove on further to the Rest House where his staff had arranged an excellent lunch with plenty of pink gins. No more was heard of this defence line, at least while I was there.

My main task, as set by the EMU, Brigade Commander, at Singapore was to plan the demolition of the road and rail communications from the Siam (Thailand) border at Sungi Pattani, right back to Butterworth. I took the unit to camp, at the Malay Volunteers drill hall at Alor Star. The States of Perlis, and Kedah, in the North, Trengganu and Kelantan in the East, and Johore, the largest State in the South adjoining Singapore, were unfederated—each had their own rulers. The Sultan of Johore was fond of fast racing cars and was reputed to have whipped people who got in his way!

British writ, was however supreme, just like in the Indian States. We were in States of Perlis and Kedah; I stayed at the dak bungalows, same style as in India. We had our own transport and started working early in the mornings. Taking detailed measurements of the road and rail bridges wasn't easy in the thick jungles, the most annoying were the sudden sharp showers, the rain would soak one to the skin within minutes, despite our rain coats.

Unlike some of the flimsy looking timber trestle railway bridges on Siam side of the border, we had substantial steel and R.C.C bridges, which needed built-in chambers for demolition of the abutments, and fabricated steel boxes for fixing gun cotton, to cut the girders. I still have the War Office notebook, on which these calculations were made. These were eventually prepared for demolition, and on the outbreak of war, these demolitions were carried out, "to perfection" in the words of an officer, as recorded in the history of the Royal Engineers.

We were, however, unable to hold and cover these obstacles by fire. The Japanese engineers who closely followed their troops, repaired the roads quickly. They also had a Corps of railway bridging units who made short shrift of our demolitions and soon had the railway system in operation.

A Christmas Party

Those were the days of plenty. I owned a large secondhand Cheverolet. We were asked to a Christmas cocktail party by our

Canadian Medical Officer and his wife who lived down town. In great form, all the young officers dressed up in their mess kit, and trooped down to my car parked in the driveway. It just refused to start. The boys pushed it down hill, but there was not a splutter from the engine. We were dismayed. Luckily Colonel ROFD Gage, the CRE, saw our predicament. He and his wife were sitting in the balcony over the porch. He hailed me and said that as his wife and he were not going out anywhere, I could take his car. Everyone heaved a sigh of relief and we all managed to pack into his Morris and off we went. The hostess certainly knew all about American cocktails with most inviting names, we tasted only a few, but by the time we left the party we were skidding on the pavement on our jingle jingle spurs, which had little serrated wheels at the ends to spur the mounts while riding. We took a wrong decision and went into a cinema. We had taken seats in the balcony, which was fairly empty. The picture was already halfway through, no one followed it. I recollect several trips to the Rest Room. Finally, we landed up at the Great World—drank innumerable Brandy—ginger ale—danced till early hours of the morning—with the hostesses telling us how much they adored us—the soldier boys! At last, the bell rang in my brain—I had the CRE's car, which had to be driven back and delivered intact. With a little fumbling I found the keyhole, turned on the engine and we were on the long journey on a narrow unlit road. It appears that either the car knew the way or some superhuman forces took charge of the wheel, till the car was parked in the CRE's drive. Next morning, I found myself tucked into my bed—very dry in the mouth and sore in the head.

Later, in spring, my section was relieved by another officer, Kashyap. On our way back we stopped for a fortnight at Kuala Lumpur, fondly called KL. Our task was to convert an exhibition centre into accommodation for the next Brigade, coming from India.

KL is a charming city, cradled in green hills. Its large buildings, starting with the railway station, stood bright and shining. One of the institutions I got to know was the then 100 year old

Selangor Club, otherwise known as the "Dog." Here members ordered Stengahs—1/2 pegs, Sukos—1/4 peg, and pecked at "Kichi Makan"—small eats with their drinks. It was a different world. The Prince of Wales on his visit in 1922 had scandalised everyone by dancing all night with a comely Eurasian girl from Ceylon. I saw the balcony overlooking the Padang (cricket ground) —where Noel Coward wrote the famous lyric *"Mad dogs and Englishmen go out in the midday Sun."*

My men supervised the Chinese workers, industrious and quick. They completed the work well within the fortnight. We rejoined our Company at Tyersal Park Singapore.

Tailpiece

At the start of the war, I was the only Indian officer at Penang, but by the end of the war the situation was reversed, as most of the British officers had been repatriated.

I never went back to Penang, but my golfing partner Brigadier Suri told me of the following incident, "When he was on the staff of Lieutenant General Sir Frank Messervey".

The Army Commander was invited to a gala party at the Runneymeade Hotel Penang, on the occasion of the handing over of the territory to the civil authority. The Governor designate and other officials were at the party hosted by Brigadier Thomas of the Somerset Light Infantry, the Sub District Commander. Ladies wore long dresses, men in Black Tie, and champagne was flowing. General Frank Messervey with Captains Ghosh and Suri sat outside in the lawn enjoying their drinks. The host got worried and came out to invite the General in. Sir Frank moved in with his two Indian officers. His invitation had read "The Army Commander and two British Officers." Later Sir Frank sarcastically addressed the gathering and said "If I were to wait a little longer I might see King Henry the Eighth at the Party!"

The Chivalrous Japanese

"Slain in battle, thou shall obtain heaven; Victorious thou shall rule the Earth, therefore stand up, you son of Kunti, resolute to fight".

The Bhagwad Gita Discourse II Sloka 37.

There was quite a party at Madras (Chennai), in August 1939 at *S.S.Dilawar* carrying Timmy's 4/19 Hyderabad, and us of the Madras Sappers. The "*Gora paltan*[2] " –Argyle and Southland Highlanders, the Punjabi's and the 40 Pathans in other ships sailed away to Singapore. It was Brigadier Paris's "EMU "Brigade" going out to defend Malaya.

On arrival at Singapore, we were shown the defences of the Naval base. We camped at Tyersal Park, an estate of the Sultan of Johore across on the mainland. I still have the Singapore Times picture of friend Amar Kashayp and self-wearing antiquated "Solar Topees". The caption accentuated the difference in pay of Amar an Indian Commissioned Officer, and myself a King's Commissioned Officer. His pay was lower but he was doing exactly the same job. This disparity in the pay of the two categories combined with the national fervour was the route cause of the formation of the Indian National Army (INA) at the fall of Singapore of which more anon.

Waiting for war, I was lucky to be posted to command my Platoon (about 60 men) at Penang. It was great having an independent command—hundreds of miles from one's commanding officer. We had a rollicking time in Malaya, Amar Kashyap whose wife had joined him at Singapore, later took over from me at Penang and stayed on with the unit. While Kumar Kochhar and I were posted back to Bangalore about the time of the sinking of the British warships the *Prince of Wales* and *HMS Repulse.*

The war in Malaya was a short one. The Japanese attacked on December 9, 1941, the day of the "Pearl Harbour;" our battle against them was most unequal, and poorly directed.

Britain had based the defence of Singapore on the Royal Air Force, but was unable to send the requisite number of planes. The Japanese first captured our airfields to gain mastery of the skies.

They had tanks while we had only the aptly called "Boys rifles" as the anti-tank weapon. Our troops fought valiantly but were unable to destroy their tanks, as we didn't have anti-tank guns. On top of it—our High Command did not agree to the preparation of alternative defence localities built in concrete anti-tank obstacles. There were no anti-tank mines to even slow down the Japanese attack—it was a most unequal fight. We occupied many a hastily chosen defensive positions but were soon outflanked by the enemy using bicycles and country boats. The road behind our men would then be blocked with the enemy all around, some on trees, taking pot shots, and throwing grenades. Our troops had little choice but to fall back. When almost the whole of Malay Peninsula was overrun, Amar Kashyap of 15 Field Company at Penang was sent back to ensure the demolition of the Johore causeway. A 70 feet wide gap was blown on January 31. The Argyle and Southland Highlanders were the last to march across it with their pipe band playing.

The Japanese troops followed closely and entered the island. They attacked all round and cut off the water supply. The British had by then lost heart. Another friend, Tony Bhagat, then commanding a 'Bombay Sappers Field Company' was holding a defensive position on the approaches to the island. He sent his British second in command, a Captain, to obtain ration—he never came back—he had scooted on a transport sailing back to India. Tony Bhagat was disgusted and joined the INA, later he fell out with their leaders.

On the 10th morning, Japanese soldiers entered the hospital, where Harbaksh was recovering. There was widespread indiscriminate shooting. The end came soon after at 6.30 pm on Sunday, February 15.

Formation of the INA

Next morning, the POWs were mustered at the Freyr Park football ground. A high stage had been erected; the prisoners were seated on the ground with the Officers in the front rows. The gathering was first addressed by Colonel Lunt as the representative of the British Government; he made a short announcement—in that he was handing the prisoners over to the Japanese. He was followed by a Japanese officer who in turn passed them on to self-styled General Mohan Singh.

Mohan Singh, who by then had acquired ADC's as a Supreme Commander, mounted on to the platform, announced the birth of a 10,000 strong Indian National Army (INA). More and more slogans were raised. As Supreme Commander, he commented that the response from the officers was weak—he decreed that they take off their badges of ranks. They will in future be treated as *jawans* (common soldiers). He was trying to assert his authority.

Those who had joined the INA wore an armband with an "F" on it. The authorities said it meant they were free, I have a sneaking doubt in my mind as to whether it stood for "F Kikan," the Japanese Intelligence Organisation, which controlled the INA.

Some of us will remember that pre-war, there used to be special cadre of high Class—tall and well built Punjabis who served in the Hong Kong & Shanghai Artillery. They helped to man the coastal and anti-aircraft guns in these colonies. Mohan Singh as Supreme Commander, one day took it into his head to inspect these gunners at Changi. He arrived there in a staff car—flag flying, ADC's and all, but was refused entry. He tried to force an entry. The Japanese sentry assaulted him. Mohan Singh was furious. Nonplused, he discussed the matter with Major Fujiwara, the head of "F Kikan." He was very politely but firmly told that "These troops have nothing to do with the INA."

The event started the arguments to the political aims of the Japanese Government—the subsequent arrest of Mohan Singh, and his deportation to Sumatra, is a separate story.

Formation of the INA had left the men confused. The Nabha Akal Infantry, under the leadership of their officers, buried their mess silver in the jungle, to join the INA. After the war when the Commanding Officer was accused of treason; he replied, "We had no intention of fighting against our army—nor were we asked to do so. We joined so that my men spent four quiet years being paid by the Japanese—instead of being shut up in a POW camp."

After 'VJ' Day, in 1945, they dug up their buried silver and had a celebratory guest night on return to Nabha!

The Japanese were the most tenacious fighters, yet when they lost the war, we discovered another side of their character. Their acute sense of discipline. Their behaviour was exemplary. Each man came forward, deposited his ammunition and his grenades in separate piles, then stacked his weapon rifle in neat pile and walked straight into the barbed wire enclosure designated as the POW camp.

General Muddy Rau (then a Lieutenant) described the scene after landing at Malay as follows: "After fighting like tigers, they after surrender, packed up, when ordered and took us on a tour of their defences, affording full co-operation." There are many stories of atrocities committed by the Japanese on their prisoners. Friend Draghalkar made a lot of song and dance about his having been hung upside down in a cage. Tony Bhagat of INA refuted it. He said, "there wasn't any room to swing a cat at the alleged torture chamber"—there is no way to get at the truth. When denied entry into British PW Camp, he wily nily returned to the Indian PW Camp. K.N. Singh tells me that Drag had admitted that he had refused to join the INA and there upon, Subhas Bose had told him to go and polish the shoes of his British Masters. In retrospect Drag had regretted his action.

All I can say that during my two years on the Tamu & Tiddim fronts, and then at General Slims HQ 14 Army, where we received minute to minute reports from the front—there was not a single report of atrocities or a case of maltreatment.

A recent story "Goldeneye's Escape" in the *Royal Engineer's Journal*[3] recounts the experiences of a Sapper along with two Gunner officers, who after Singapore's capitulation in 1942 had sailed away in a dingy. They almost got to Java—14 days out they were almost at Sundra Straits—between Sumatra and Java; when they were intercepted by the Japanese Navy.

At first, a cruiser instructed them to land at an island—but they had disobeyed and carried on. For the second time, within 20 minutes, a voice boomed through a megaphone in English—but not so agreeably polished, "Stop or we'll shoot."

"We stopped, I mounted the Jacob's ladder and—as my feet touched the deck—a rating slapped me lustily across the face. When the other two appeared expected to be hit too, but my tormenter received a resounding clout across his face by his officer for having struck me without orders.

We were cursed roundly for having disobeyed orders. We were on the Admiral's flagship *"Asigara."* A Lieutenant Commander who had been at Oxford, interrogated the three of us. After ten minutes of the first interview, we were ordered to bow to the Captain and depart. On the way back, my blister was treated and we received our meal of rice and fish, followed by a tin of oranges and sweet coffee.

"Later, a table was set up for us, we were issued knives and forks from the ship's store—an audience turned up to see the use of the strange implements. We received regular meals. We were asked to give our word that we would not interfere with the working of the ship; after which we were allowed to stay on the deck. It was a case of old world courtesy and chivalry. In any case none of us had the least idea as to how one goes about sabotaging a warship."

My neighbour General Tindi Sharma told me, how when as a subaltern, he was given a working party of some 150 Japanese to load their ship, for the journey back to India from Surabayia. He was greatly impressed by their dedication to duty.

The rest of the story is about the experiences of three, ex-students of Government College Lahore—Harbaksh, Amar and Dara—in Malaya. The first two were prisoners at Kluang Airfield and Dara, the famous hockey player was an INA (Indian National Army) official, in charge of the Indian POWs.

Harbaksh narrated to me his experiences as a POW, as we sat together at a birthday dinner party. It was so different as to what one normally hears, that I sat down and put it on my computer, and sent him a print for approval. He told me that he was writing a book. This indeed is now published and titled, *"In the Line of Duty—A Soldier Remembers"*

What I write now is the soul of the story, with the permission of his wife, now that the great soldier is no more.

A sizeable number of the Indian POWs, approximately 1,500 men comprising the Jind Infantry under Colonel Gurbaksh Singh, 40 men of the 5/11 Sikhs under Harbaksh, and men of the 5/12 FF Regiment, had joined the ranks of the prisoners. They were grouped together under Colonel Gurbaksh Singh of Jind State forces and worked at the maintenance of Kluang Airfield, for the Japanese Air Force.

Harbaksh Singh of the 5/11 Sikhs, the Commandants' younger brother, and Amar Kashyap of 15 Field Company, who joined them later: both were senior Indian Commissioned Officers, they had also suffered discrimination in pay and allowances between King's Commissioned and Indian Commissioned Officers but they did not join the INA.

Amar Kashyap had been separated from his men and was severely beaten up, to make him change his mind. So much so that his spectacles were broken, he went about for sometime half blind. The Commandant of his camp, on learning that he was an engineer and had built airfields, picked him out and sent him to join the POWs at Kluang Airfield.

The prisoners there, left to fend for themselves in the jungle, under the command of Colonel Gurbaksh Singh organised and spent the rest of the war at the airfield.

Captain Harbaksh Singh, as the younger brother of the Commandant, with great courage made several forays from the camp, to the Indian communities settled in the neighbourhood of their camp to get help for the prisoners and particularly the sick in the hospitals. They formed a small circle of friends in midst of gloom and deprivation.

Captain Dara, the famous hockey player, Harbaksh's class-fellow from the Government College, Lahore, was an active member of this circle.

From his HQ at Kuala Lumpur, he as the Commander of all the INA, POWs, helped them with supplies of essential medical stores through the Indian Independence League. The three years in the camp gave them an exposure to the psyche of the Japanese. They were callous, but also chivalrous to the core. Both my friends are no more, I have recorded what I was told, as accurately as possible.

A great difference in Japanese attitude—was their complete lack of fear of death. They were real believers in the philosophy of the Gita. Brought up on the glory of "*Hara Kiri*," the practice of ripping open their stomach—in event of falling from grace: and displayed unwavering allegiance to their Emperor. They were therefore, quite understandably, on better terms with the POWs—than the INA soldiers. There being no greater ignominy for them than to turn against the Emperor, to fight against their brethren.

A friend, Ambassador, Khurana, originally of 8 Punjab, who was taken prisoner within a week of his arrival at Singapore had married a young lady and worked for the INA, told me that the Japanese never saluted INA officers.

The Japanese disregard for pain—possibly the Shinto gospel, analogous to our Vedanta teachings—made them more callous. A section of the Japanese Army detailed to look after the security of the camp received a piglet, as its weekly meat ration. The cook would cut off a leg each day keeping the animal alive, by applying a tourniquet. This cutting off the limbs of animals, without compunction abhorred us but in the absence of refrigeration, it was the only course open to them.

Another example of callousness was the way Japanese recruits were shot at with live ammunition during field craft training to teach them to keep their heads down. Even their doctors were very callous—instead of giving injections to the men individually, they would order a whole unit to parade and would inject everyone with all the three serums—cholera, tetanus and TAB; at one go. They were then given leave for 24 hours to recover.

The most obnoxious was the "Stool Test." They would order fifty men at a time to be lined up, with trousers down. The doctor would move along the line, insert a glass rod into the anus of each soldier, take a sample, smear it on the glass strip held by an orderly; there was no sense of shame! A sure quick way—but callous.

There was, of course, the other side of compassion. Major Sato, the Camp Commandant was a serious student of "Vedanta". He had a library of books by Swami Vivekanand, and practiced yogic concentration "*Isht*" at a temple in his quarters. Wanting to extend his knowledge he befriended Amar who had not only been practicing Yoga but had also delved into the mysteries of hypnosis. I remember the meetings at their Chouburji home at Lahore. The family would gather around a mahogany table and would call up spirits of the departed. I had stopped attending such seances—as I was unhappy at disturbing the spirits of departed.

Another interesting person on the Airbase was Captain Cani, an ex-Professor of English at the Tokyo University. Wanting to keep up his knowledge, he spent quite some time with the POWs practising his English. A fighter pilot, he had volunteered for "Kamikaze," a suicidal attack (literally to be a divine wind). He was the idol of all the pilots on the station.

Harbaksh once asked him about his family. He took out a photograph of his wife and children from his purse and showed it to him adding, "I have entrusted them to the nation and have no more attachment."

When, at last, orders were received at the base for a Kamikaze attack on the enemy fleet, seven planes were lined up, garlanded and took off after a solemn ceremony. The wireless communication centre then became the focus of interest. At first, it was reported that the flight could not locate the enemy fleet due to dense cloud cover. Eventually Cani found a gap—and the seven hurled themselves at the Warships—Theirs was a Do and Die mission.

The war ended when Atomic bombs were dropped on Hiroshima and Nagasaki. The prisoners were liberated. A British

War Crimes Commission, under a Brigadier, questioned the POWs. They were asked to identify and file reports against their Japanese jailers—to enable them to be tried as criminals. The Indians at this camp under the leadership of Colonel Gurbaksh Singh decided not to submit any complaints.

The Japanese Commander, in his full uniform, came to Colonel Gurbaksh Singh in his staff car—flag flying along with his staff. He bowed before him saying, "You are a much greater man than I." He handed him a list of his weapons, asked him to move his men to the barracks and headed off to the POW cage.

He had hardly gone a hundred meters—when he retraced his footsteps. Remembering the Sikh religious injunctions against tobacco, he requested permission to remove his obnoxious cigarette case, which he had inadvertently left in the staff car.

Post script

Slender is the Thread between abject misery, and upright existence

The Times of London reported that a new film, "To End All Wars" based on the story of Captain Earnest Gordon of our "Gora Paltan" the Argyle and Southland Highlanders, is to be shown at the forthcoming Film Festival at Cannes. He claims that 100,000 POWs and local people died while building the Siam-Burma railway link.

With a slender thread hung the fate of our 1,500 men under Colonel Gurbaksh Singh of the above story. They were first destined to be sent to Rabal Islands. As the ship to take them failed to turn up, they were loaded on cattle trucks to work on the "Death railway." It was by sheer luck that they were off loaded at Kluang by the Air Force to work at the Airbase and escaped the worst.

Talking to Brigadier Litty Malhotra, the winner of the bottle of whisky for having hoisted the flag at the Government House at the liberation of Rangoon, I learnt that the two thousand or so British POWs there, were in a poor state, suffering from deprivation. A Japanese officer, captured by him and who was escaping disguised as a Burmese, was not so well off either.

Physically the best fed of all was Shah Nawaz of the INA who surrendered to his battalion. The irony of fate was that he was awarded the gift of a *jagir* and made a Minister.

General Bhosle of INA ,who was at their HQ came back to India with a wad of high denomination Thai currency notes which he gave to General Shiv Dev Verma for safe custody when he met him on the train from Calcutta. Bhosle, too became a Minister and was being appointed the Lieutenant Governor of Himachal Pradesh, but Army Chief General Carriappa intervened.

Such is the irony of fate; it is not always the most deserving who reap the benefits.

Notes

1. A device for blowing a hole in the ground, to take a large explosive charge.
2. British battalion
3. R.E.Vol 112, No. 1

4

War in the Desert

Fighting in desert in North Africa was a strange experience for our men. There were no trees except near the coast, and no vegetation, but the desert was friendly. The 4 Indian Division was the first to arrive at Mersa Metruh—a small railhead in Egypt, about 200 miles from the Delta and 120 miles from the Libyan frontier.

The Italians, had occupied Libya in 1911 on the pretext to go to war with the Turkish Empire. In 1914, the Arabs rose in revolt and pushed them out except in the coastal towns. In the 1920's, the Senussis were herded into camps, and the Italians took charge. The northern part of the country is called Cyrenaica which is a bit confusing.

The Italians had a large garrison of 14,000 officers and 3,27,000 men occupying the colony, Libya. A country more than twice the size of Egypt, but with a small population of only two millions. The 4 Indian Division, the first to arrive in the Middle East, camped at Mersa Metruh. The 5, 8 and the 10 Divisions followed later, to take part in war in the desert.

In the summer of 1940 when the Germans broke through the French defences, Mussolini declared war. Our 4 Div. moved into a defensive position in the desert and then on the night of December 7, its 6 Brigade moved out of camp, to take part in a training exercise.

This slid into real war. The Brigade spearheaded an attack on the Italian defensive position at Nibewa, and captured it within the hour. This started a run by the Italians. Our 7 Armoured Division "Desert Rats" trained by an extraordinary Sapper—Gen. Horrocks—followed up and took Siddi Barrani and the ports of Tobruk and Benghazi, the capital of Cyrenaica. Advancing rapidly, our troops were in Tripoli, colonized by hard working Italian peasants, who with great labour had developed farms and orchards of fruit trees. They lived in identical white bungalows labeled "Ente Colonizzazione Libya"

By December 15, all enemy troops had been driven from Egypt, the Italians had withdrawn into the defences of Bardia. In the rapid advance of 400 miles westwards our troops captured 130,000 Italians including seven Generals, 1,300 guns and 400 tanks.

No Italians were left in Egypt except those wending their way to the prison cages—enroute to camps in India. The operations went so well that the 4 Indian Division was relieved by the Australians, and sent to dislodge the Italians from Abyssinia.

That our offensive had halted, was not apparent to the Italians—they implored Hitler for help. He sent Rommel, one of his most competent Generals. He was also allotted two Panzer Divisions. Rommel had gained great fame as Commander of the 7 Panzer Division in France, and had almost become a national hero.

Rommel, on arrival at Tripoli on February 12, straightaway carried out a personal reconnaissance by air. Without waiting for the arrival of his two Divisions, he struck on March 31. It developed into a full-fledged attack. He re-took Benghazi, along with its stores, captured most of Cyrenaica except Tobruk, which was held by the Australians and 18 Indian cavalry. He was always in the front—with the advanced guard—lived on the simple fare of the troops. In a short period of eleven days, he drove us back to the borders of Egypt. He was fighting against the British forces reduced in number and quality. The 7 Armoured Div (Desert Rats) had been withdrawn for refit and had been replaced by the newly arrived armoured units. The 3 Indian Motor Brigade had moved up to take the place of 4 Indian Division, that was guarding the aerodrome of Tobruk. It was the only mobile force; but it had only soft vehicles and was armed with 2 pounder ineffective anti-tank guns (Popguns). Tobruk was held, but the 3 Indian Motor Brigade was badly mauled.

Lieut. General Tuker—the famous Commander of our 4 Indian Division—in his book *"Approach to Battle"*, says, "The 2 pdr Anti-tank gun nearly lost the British the Middle East and the war." To make up for the 2 pdr popgun– "we ferreted out in Cairo a number of spigot mortars which were thunderously most destructive and gave them to our Sappers. These witch-pots propelled with fair accuracy up to 500 yards a whacking great mortar bomb fixed on the end of a stick like huge lolly, which created hell and havoc—where it landed". Luckily, the 8 pdr arrived in May 1942.

On the administrative side, all petrol was supplied in flimsy kerosene tins—half leaked enroute. The only advantage was that these could be hacked into half for brewing tea!

A squadron of the 2 Royal Lancers under Major Rajinderasinji[1], with a few British and Australians, broke out into the desert and reached Egypt. Rajendrasinji won a well-deserved D.S.O.

Rommel's tanks were halted as they ran out of petrol, unknown to the Italians who were responsible for supplies in this theatre, their supply ships were being sunk—as the British having broken their radio code were thus able to sink them.

Battle Axe – In May 1941, Churchill who wanted General Wavell to give an early reply to Rommel's attack, sent fresh supplies of 238 tanks through the 'Tiger 'convoy which was successfully taken through the Mediterranean. Churchill expected an immediate offensive whilst Wavell wanted time for overhaul, camouflage painting, etc. This delay in the start of British offensive by the Eighth Army started the misunderstanding between the P.M. and Wavell. The new arrivals included just over a hundred Crusaders and 135 Matildas. The Crusader tank had a top speed of 15-20 mph and radius of action of 80-100 miles. It could not be matched with the Matilda, the "I" (Infantry tank) designed to keep pace with the Infantry walking into battle, moved at a speed of only five miles per hour. These were thus no matches for the much faster and maneuverable Panzers.

They also had their teething troubles, and the 7 Armoured Division that had been without tanks for four months took time to train their tank crew.

Battle-axe, opened on June 15 and by the June 17, every thing went wrong. The 7 Armoured Division was detailed to cover the desert flank. The Matildas attacked the Italians who held posts at the top of Halfaya (called Hellfire) Pass. These tanks which mounted the 2-pounder popgun whose shots dropped off the enemy tanks like peas from a peashooter were torn to bits by the German 88-mm anti-aircraft guns used in the anti-tank role. Rommel had experience of their use in anti-tank role in France and had personally ordered their allocation for defence of Halfaya Pass, half of the Matildas, were destroyed, the rest had to fall back. The British Anti-Aircraft guns, being of riveted construction could not be used in this role as the rivets would start popping out in the horizontal mode! The operation was called off. Churchill was very distraught at the failure. General Auchinleck of the Indian Army replaced Wavell. He, on receipt of 300 American—fast 13.5 ton Stuart light tanks—called "Honey's" under the "Lend Lease Agreement" launched the Crusader Campaign on November 18. These mounted a 37 mm gun which fired 'capped' improved ammunition.

The aim was to destroy Rommel, and relieve the besieged garrison of Tobruk, which was surrounded by three Italian Divisions, plus a fourth in reserve. For the first time, over 600 tanks were hurled into the battle.

The crack 4 Indian Division was back from Eritrea, having finished off the Italian Empire in Abyssinia in conjunction with 5 Indian Division. There were unexpected rains, but the troops were in high spirits and crossed the frontier, to advance towards Tobruk 90 miles away. It went well for the first three days. Sidi Rezegh airfield, only 12 miles away from the perimeter of Tobruk, was captured.

Rommel, who was away, celebrating his birthday with his wife at Rome hurried back. He attacked and made a deep raid into the British rear areas. It was during this "putsch" that he took personnel of 17 Field Ambulance of 4 Indian Division, as prisoners.

Captain Ved Parkash of the Army Medical Corps told me the story of the encounter, as under:

"A German attack developed, they broke through the South African Brigade. Our Air Force started dive bombing and strafing an RAF pilot, crash landed only 300 yards from our reception treatment penthouse, he miraculously escaped, fired his very pistol into the petrol tank of his crashed plane to set it ablaze. Casualties started coming in from both sides including two Italian doctors who were not seriously wounded. After taking some mugs of hot tea and cigarettes they offered to help; detailed them to look after their own men.

Soon after we got involved in artillery cross fire and had to take to our slit trenches. Our Jemadar, Chander Bhan Misra lay amongst the medical panniers, and kept on saying *"Aaj to nahin bachte sahib"* (we will not survive today). The German tanks came, an officer with a luger in his hand ordered us to get out of the trenches. He asked the name of the unit, and told me as the senior Indian officer, to take charge and ensure that no one escaped. A detachment of Italian soldiers, under a German N.C.O (Non Commissioned Officer), was detailed to guard us. It was the month of November and the nights were extremely cold. To keep warm, the men took extra clothing from the stores and were wearing three shirts each. The two Italian doctors helped us; they brought blankets, biscuits and drinks from the mess lorry. At about midnight when we went to change, the blood transfusion bottles in the blacked out wards, a British officer with maps dangling around his waist walked in, saying he had a battery of guns with him and had lost his way. Major Arid, the surgeon from the attached British Surgical team, told him, "We are on boundary post 57 on Libyan border. It is in enemy hands and we are POWs; the guard has just gone outside." On being told that, he muttered, "how embarrassing" and just walked away.

Next morning, the Second in Command—Major Reid and the Roman Catholic Padre asked permission to go to their tents to collect their shaving kits, and quietly escaped in a truck from the other side. The Germans fired at them but they got away.

The German officer was furious; he held me responsible and wanted to shoot me. My Jemadar interceded on my behalf, but to no avail. Thinking quickly, I pleaded "Look, both officers who escaped were white, they would not have done it if I was also white; they don't take any notice of us colonials." He let me off with a warning. Further, as a concession he allowed the staff to prepare breakfast.

Soon after, a column of about 80 armoured vehicles and tanks surrounded us. Immediately our own guns started shooting at them. Their Commander, Col. Stephen,[2] was hit by a shell in the chest, and was brought in for treatment. The Major of the Surgical unit was a very able surgeon—who later became famous and was Knighted after the war—operated on him. Despite his best efforts the Colonel died, and we buried him in a common grave.

Next afternoon, we were ordered to get spruced up, the great General Rommel himself was to visit us. He arrived like a whirlwind, with his personal escort of three tanks, an armoured car and a staff car. We were lined up; he came, shook hands and asked if we needed any help. I said, "General, if your tanks were a bit further away from the hospital—our guns would not shell the area, and we could work in peace." Within minutes the tanks moved off. A German medical unit with a well-equipped operation van arrived. The Major spoke English; he had been a practising doctor at Port Said. Soon after a truckload of 'sowars' of the Central India Horse Regiment, strayed near us, finding us prisoners they zig-zagged away.

That evening when we were all huddled into a small 180-pound tent, a very young German officer, hardly out of his teens, trundled up to the tent in his tank and asked about Col. Stephen. We gave him the sad news and handed over his ring and personal belongings. He promised to return in the morning. To be ready for the visit, Phillip Daniel, the Protestant padre re-arranged the grave and put a bottle with Stephen's name next to his body.

One of our planes flew over us; we gave it the recognition signal—waving our hats. Two South African armoured cars strayed in, and gave us the location of the gap in the minefield. With only a few Italians left to guard us, we decided to break out. There was not enough transport, our British N.C.O, who had seen vehicles nearby, brought in a staff car and a truck. Our Italian guards got wind of our plan and requested that they may also be rescued. We agreed and along with our guards got away at first light next day.

Once through the minefield, we met with our D.R. (dispatch rider) who guided us back. Meeting our sister Field Ambulance with Bhaggu Khurana[3] was great. He got me a razor from the Operation Theater to take off the beard! Richard Dimbleby—of the B.B.C came to record the story of the escape. Within a fortnight Ved Parkash could be heard on the BBC news. His parents living in a remote village—Pind Dadan Khan—received a letter from the A.I.R. informing them the date and time of the transmission. The doctor owned the only radio set in the village. They went to the dispensary, well in time to hear the broadcast. After listening to it—his father emotionally asked the doctor to please switch it on again."

There was a clash of armour at Sidi Rezegh airfield only 12 miles from the perimeter of Tobruk. Fighting continued through November without any tangible results. On November 23, "Totensontag" when the Germans remember the day in honour of their dead, our General Cunningham lost his nerve. Auchinleck came forward, relieved him and put Ritchie in command. He ordered the 4 Indian Division and the 7 Armoured Division to cut off Rommel's supply lines—Rommel drew back.

On December 4, the 4 Indian Division along with the New Zealanders advanced to Bardia and relieved the garrison of Tobruk on December 11. The 4 Indian Division went on to capture Benghazi. The development of the port started and 14,000 prisoners were taken at a cost of only 500 casualties. Unbelievable but true, Rommel received a consignment of 30 tanks at Benghazi, before the port was closed.

Luftwaffe's increased presence, and with Royal Navy's misfortunes at sea, Rommel received six shiploads of tanks at Tripoli on January 5. At about the same time, the British

Armoured Brigade also received their first consignment of Grant tanks. With new tanks, Rommel again advanced Eastwards on January 21. Two days later, 4 Indian Division was obliged to vacate Benghazi despite political pressure from Churchill; it was not possible to hold on to it.

Rommel entered Benghazi, captured stocks of materials. He was promoted to Colonel General. In January 1942, there were a lot of advances and retreats, nothing spectacular was achieved by either side. The desert was alive with enemy columns. Lieutenant Mehbub Singh, the transport officer of the Rajputana Rifles, showed praiseworthy ingenuity. He drew petrol for his vehicles at the same time as the Germans were replenishing at the other end[4].

GAZALA- To relieve pressure on Malta, Churchill pressed Auchinleck to attack, using the 5 Indian Division which had arrived back after clearing the Italians in Abyssinia. Auchinleck after a four months delay planned to attack on May 20. He was pre-empted by Rommel who sent his 21 and the 15 Panzer Divisions along with the Italian Ariete Division in a flanking move to the south of Bir Hacheim, outflanking the minefields. The Germans attacked on the moonlit night of May 26/27. The battle commenced at 9 p.m. with an attack on the rear of our defences, held by the 3 Indian Motor Brigade at Bir Hacheim.

The 2 Field Regiment of Indian Artillery had been sent up in support of the 3 Indian Motor Brigade. My friend, late General P. P. Kumaramangalam (Kay), was at the time, commanding the Sikh Battery of the 2 Field. I repeat his story as he told me:-

"The regiment moved up to the position about 4 p.m. They had been promised an armoured regiment—which did not turn up. On arrival I told every one to dig, as we were on the possible route of the enemy. At about 9 p.m., I told them to try and get as much sleep as possible. My battery O.P officer, Captain Frisbee kept on reporting a lot of enemy movement on his left. I reported it to the Brigade H.Q.—who told me not to worry—these are our own South African Armoured cars.

In the morning, to my horror I found that it was German lager soft vehicles who were replenishing their tanks. I immediately opened fire, the soft vehicles withdrew, but the tanks attacked my gun position. One of the guns under Havildar Modam Singh kept on firing A.P. shells at point blank range, and caused havoc. In the confused fighting I got into an armoured car of the 18 Cavalry and was able to cover some guns to knock off several attacking tanks by direct firing on open sights till over run physically. Havildar Modam Singh got an I.O.M and Kay won a D.S.O. He and a number of our Cavalry officers including Sahibzada Yaqub Khan—Pakistan's Foreign Minister became Prisoners of War. Kay only learnt of the award after he got back from the captivity. The 3 Indian Motor Brigade almost ceased to exist.

Rommel's Panzers attacked Tobruk on June 20. The Eastern perimeter was being held by 11 Indian Brigade, the main attack fell on a single battalion of the Marhattas who were holding a three miles of unmined front, with only Pounder Guns and a detachment of 25-pounders in the Anti-Tank role. They fought hard, their carriers counter-attacked valiantly. The Germans attacked with the help of the Italian forces under concentrated artillery bombardment, supplemented with 400 tones of bombs by the Luftwaffe. They destroyed 260 tanks. Rommel entered Tobruk on June 21. He was greatly helped in his further advance—by the capture of 1,400 tons of fuel—2000 vehicles and 5000 tons of provisions. He was promoted to Field Marshal. Erwin Rommel, thus became the youngest F.M. in the German Army. He was so busy that he forgot to wear his new badges of rank till F.M. Kesserling came and gave him his spare set.

The news of the loss of Tobruk reached America, when Churchill was in conference with President Roosevelt—whose immediate reaction was to ask, "What can we do to help?" Churchill asked for three hundred Sherman tanks. Within a short time not only the Shermans, 100 S.P. (Self Propelled) guns and fighters were withdrawn from American Divisions and put on board—six of the fastest ships for Egypt.

This was followed up by the loss of Mersa Matruh where my friend Lieutenant Shamsher Singh had the unenviable job of destroying NAAFI stores, choicest whiskies and chocolates and all. At the end of it, he and Major Ajit Singh of the Supply Corps were also taken prisoner. They joined a very distinguished company—including two future Presidents of Pakistan—but that is a separate story.

General Shiv Dev Verma, then a Lieut. Colonel on the staff of the 10 Indian Division, remembered seeing the fire works display at Mersa Matruh. He drove back so furiously that even Girdhari, his driver had to say *"Dheeray Chalo Sahib"*—so did the rest of the Division. The situation was grave on June 25, General Auchinleck moved up and took over personal command of the VIIIth Army. He planned and started manning the fall defensive position at Al Alamein.

Rommel's troops reached the gates of Egypt. Rommel wrote to his wife Lue on June 30, that he was only 100 miles from Alexandria; by the evening it was only 60 miles. Mussolini got ready to enter Cairo. He landed at Derna—on a white Charger that was to be flown in on a separate aircraft. The way was clear, our Headquarters—burnt their files—it was Ash Sunday! The Royal Navy started sailing away from Alexandria.

Al Alamein It is not generally appreciated that there were several battles of Al Alamein. This first and the crucial battle that held the German onslaught, took place on Wednesday, July 1 under General Auchinleck. Four strong points—called "boxes" were held between the coast and the Qattara depression (an impassable salt marsh) a distance of 35 miles. The strongest position on the coast was held by the 1 South African Division, two were held by Indian Brigades and one by the 6 New Zealand Brigade. Little known is the gallant defense of "Dier Ei Shein" by the 18 Brigade of the 8 Indian Division on July 1 and 2—as brought out by Captain Liddel Hart's book that saved Egypt. Their name deserves to be written in letters of GOLD. This gallant 18 Indian Infantry Brigade of the 8 Indian Division—ceased to exist having fought to the last survivor and was not re-raised.

In his daily letter to his wife, Lue, on July 4, Rommel wrote, "Things are unfortunately not going as we would like. The resistance is too great and we are exhausted."

Churchill arrived at Cairo on August 4, and in consultation with F.M. Smuts of the South African Army, appointed General Montgomery to command the VIII Army. Monty made his H.Q. more comfortable by moving to the seaside. A careful planner, he took two months to launch the second battle of Al Alamein. At 21.40 hrs on October 23, one thousand guns fired a barrage and 500 Shermans with their 75 mm gun, went into action for the final battle.

The magnificent forward drive of the Australians—Operation "Super charge" commenced at 1 a.m. on November 2, 1942 under a barrage fired by 300 guns, our troops broke through. After this action there was never a defeat. One of the notable actions was by a 20-year-old Lieut. Nand Lal Kapur of the Rajputana Rifles, who led a successful attack on the Ruweisat ridge. At the end of it he stood up to have a look and was surprised when a large body of Germans—thinking that he had a large force behind him—gave themselves up as POWs. He won an M.C.

The acting German Commander General Stumme died of heart attack. General Thoma was captured. Rommel returned from sick leave to save the battle, but was forced to accept defeat on November 3. Thereafter, he remained short of ammunition and supplies, and lost Tobruk as a base, on November 13. Rommel's end came due to plentiful arms supplies from America and sinking of his supply ships by the Royal Navy and the RAF. Rommel a sick man, fought his last battle at Medenine on March 6, 1943, and then proceeded on sick leave, handing over the command to General Von Arnheim.

The next battle was at the crossing of the Mareth line—a 22 mile long obstacle some 200 feet wide and 20 feet deep. The Madras Sappers under Major Cameron, not only prepared facines for crossing the obstacle but also set a good example of steadiness under fire.

Operation Torch

The Americans launched Operation Torch. Their troops landed on the North Coast of Algeria at Algiers on November 8, and other ports on Nov 11 and 12, 1942. They moved rapidly and took over airfields. Two British parachute companies supported by commandos landed.

American Parachutists dropped at Tebassa on Nov 15 and moved to Gafsa, in the south of Tunisia. The the VIII Army, which included the Indian Divisions came under General Eisenhower, whose main problem was to deal with the Free French.

Whatever else, when it came to punching a way through the German defenses, the job was accomplished by the 4 Indian Division. The 4 Battalion of the Rajputana Rifles led the attack. CQMH Chelu Ram, without caring for his own safety, with a handful of rifle men attacked, and routed a strongly held defensive position. He was awarded a Victoria Cross.

The Germans surrendered to H.Q. 4 Indian Division on May 11/12, at Tunis. The Caravan of the German General Officer Commanding Von Arnheim is now with H.Q 4 Indian Division. My only lament is that the battlewagon which had been presented to India to commemorate the occasion, is not lodged in National Indian Army Museum.

Havildar Sampangiraj in Abyssinia, 1941

After Keren, the advance continued to Massawa, the seaport about 50 miles away. Company Havildar Major Sampangiraj of 12 Field Company who was sent by his C.O. to deliver a message to the second in command, was detailed by the latter to help the Royal Sussex Regiment in their advance to capture Massawa.

I reproduce his story, as recorded by him for me, in the memory of this great Madras Sapper who rose from a sapper to command of a Field Company.

"It was on April 7, about 2 o'clock, when the Sussex vanguard Commander asked me to clear a gap in the minefield for his carriers and troops to go through. Meanwhile, they withdrew and deployed in the cover of scrubs and broken ground. It was a bright and sunny afternoon. I was alone in the no man's land, in front was an enemy minefield more than 200 yards deep. From thence onwards were the perimeter wire—a thick wire fence—enemy's pillboxes, weapon pits and anti-tank guns covering the approaches. It was a very tense and anxious moment."

"I could see the enemy movements near the pillboxes. I knelt down and started to disarm and lift the mines. Wondering all the while as to why the enemy only a few hundred yards away was so quiet, while I was lifting their well laid mines with anti lifting devices. Perhaps it was the afternoon siesta time; I managed to make good progress single handed."

"Soon after, the whole front suddenly came to life with a roar of guns. Sheets of machine-gun tracer bullets crisis-crossed and tore through the air. Grenades and mortar bombs burst all round. The anti-tank guns made a shattering din, some mines started exploding. Looking back I saw some of our vehicles on fire. I was now approaching the enemy's perimeter wire. Most of my body was covered with smoke dust and mud. At last about 7 o'clock in the evening, still bright in the African sun, I had completed the clearance of the gap. The mines had been placed on the sides to mark the gap. The high wire fence I found to be electrified. I turned around and crawled back to where our troops were. Three senior officers who had been watching with their binoculars, shook my hand in great appreciation."

"Captain Taylor, the 2 in Command offered me a mug of water. The same night I guided the attacking troops through the gap, and blew the wire fence on the enemy side of the obstacle belt."

"When I reported back to my Company Headquarters, the Subedar Company was furious as to why I had not reported back after delivering the message to Capt. Taylor. The Commanding Officer gave me a severe reprimand for dereliction of duty. I did not blame the Sussex Officer for assigning me the mine-lifting task. Whilst I was being marched out, a Brigade dispatch rider arrived with a message for the O.C."

"The Subedar Company was called back into the office. When he came out he told me to pack my bags and proceed immediately to the Indian Reinforcement Camp at Cairo, on 7 days leave. I was shocked at the double punishment, of also being sent away from the Company, which I had served so loyally."

"After a long and tedious journey by sea, rail and road, when I reached Cairo, I found an operational immediate signal waiting for me, which read, "To Jemadar Sampangiraj, Congratulations on the immediate award of I.O.M—The Indian Order of Merit" carries with it a lifetime monetary award. The C.O. had arranged to give me an antedated promotion to the Viceroy Commissioned rank, which conferred a higher allowance for me and my family for life. A very generous act, on his part".

Shri Guru Granth Sahib

By Col. Hari C Kapur

Our troops fought in several lands, but I wonder how many of us, have ever heard of the doings of our famous 4/13th Frontier Force Rifles—the Piffers—on the remote island of Castelo Rezoi—(Castle Rosso). I happened to be sitting next to Col Hari Kapur at a dinner.

He had joined this famous regiment in Egypt during the Second World War. The officer shortage was acute. After coming out of the Officer's Training School, he had spent hardly a few months at the training centre at Abottabad when he was put incharge of a posse of men, Pathans, Dogras, Punjabi, Mussalmans and Sikhs—the four classes—and detailed to take them to join the regiment in the western desert.

Eventually, he and his party reached Cairo—GHQ. The Headquarters said they were not sure where exactly the battalion was but gave him a compass bearing and told him to go on till he found them. This was his first introduction to the famous battleground—a vast expanse of desert—bare and treeless. Luckily, after a few hours, he started seeing the black hose tops, the Piffers wore black badges and stockings.

The regiment was badly mauled during Rommel's attack. It was withdrawn for rest, refit and training and finally landed up at Cyprus. Their escapade to Casle Rosso took place a few months later.

The incident narrated in this episode is connected with the famous error of judgement of Mr. Winston Churchill, who decided to intervene in Greece. The 6 British armoured Brigade, the 7 Australian Division with New Zealanders, and Polish Brigades started moving on March 3, 1941, to join the Greeks and the Yugoslavs. The Germans launched massive attacks, the Greeks despite British help, withdrew. The British forces evacuation ensued, unlike at Dunkirk, the German Airforce was in command of the Air. Working from small ports and beaches the Royal Navy carried out the rescue operations on the night of April 24.

It was in this scenario that our famous 4/13 F.F. Rifles, was sent forward to occupy Castle Rosso—a small island only about seven miles long and three mile wide off the coast of Turkey—an insignificant island which is not even shown in normal atlas. It lay in the Diodecanese South-east and is not far from Rhodes.

The ships were all busy getting the Aussies and the British troops out of Crete. So the regiment embarked on M.T.B's—small boats about 35 feet long, powered by a V-8 engine. Each of these boats had an Anti-Aircraft gun mounted in the front. With great difficulty the men managed to get into the armada of about 30 boats. The M.T.Bs were mighty uncomfortable, moved at a speed of about six knots; and were attacked by the German Stukas—in spite of being escorted by two destroyers. Luckily none were sunk. To get a little respite, the convoy was moved into a Turkish fiord for the day time and renewed the forward journey at nightfall. Turkey was neutral in the war at that time; their commander therefore allowed the halt for 24 hours for repair as per the maritime custom.

They were glad to reach their destination—a beautiful island, lush green in a blue sea. Gentle breezes blew in the excellent spring weather. At the highest point on the island, towards the Turkish coast about three miles away was an Anti-Aircraft gun position manned round the clock by British gunners assisted by West Africans. The German Luftwaffe was extremely active, the Stukas would come roaring past the gun position. Hari remembers one gunner being cut into half by the burst of bullets.

The battalion soon got down to occupying a defensive position, they dug trenches on the high ground around the perimeter of the island. The enemy airforce was active, the battalion inspite of being well dug in lost 12 men. The British Medical Officer (M.O.), suffered shell shock one day, and he ran down through the perimeter in to the sea. He was restrained with great difficulty. A Sunderland flying boat came to pick him up.It was a beautiful sight on the blue water. Unfortunately the boat carrying the doctor and other causalties for evacuation rammed into and punctured one of its floats. The graceful flying boat slowly sank into the sea. Another that followed a few days later and brought them mail and a new doctor.

One of the problems was bathing. Officers went down to the beach, which was infested with sharks and a species of sea ticks. Hari unfortunately got infected. The doctor applied spirit and had to shave him all over the body with a razor—he spent three days very uncomfortable.

The town lay in the middle of the island, a prosperous community. The residents had gone away leaving their homes intact, neat and clean. The only movement was by robbers who came from the Turkish coast to steal carpets from the deserted homes. Hari recalls how during patrols through the town they would be tempted to have a lie down on a spring bed, after months of sleeping on the ground. The atmosphere was so calm, when not being strafed, that the Subedar Major Wasan Singh established a small Gurdawara in one of the houses in the village. On Sundays and holidays the Sikh Company with Wasan Singh leading, the Guru Granth Sahib on his head, would meander down to village to their Gurdwara. It was a prefect example of peace and contentment due to their deep religious belief. There was never any panic—and believe it or not never during the month they were on the island, there was no attack on this small religious procession to or from its place of worship.

When Rhodes fell to the Germans, our troops came drifting back on rafts, and planks. The position was untenable, suddenly orders were received to abandon Castle Rosso. They embarked, with personal arms only, on three destroyers—one British, one French and one Italian. They said good bye to the lovely island at 0003 a.m. and were back in Cyprus in good time for breakfast.

Italy

The quest for a second front, to help the Russians, led to the Italian campaign—an attack at the soft under belly of Axis resistance. It opened with attack and capture of Sicily. General Eisenhower of the American Army was in overall command. The British VIII Army, under General Montgomery landed in Sicily on July 10, 1943. Two of our Battalions—the 3/10 Baluch and the 3/12 Frontier Force—supported the landings, but were not involved in serious fighting. The Germans were pushed out by August 17. Mussolini, the dictator of Italy, fell from power on July 25. This campaign lasted only 38 days.

Italy Invaded

The Allied Armies crossed the Straits of Messina into Italy, on September 3. Surrender of Italy was arranged secretly on the same day, but it was not announced till the Anglo-American Vth Army had landed at Salerno south of Naples (Napoli) on the 8 September. It was then announced by the B.B.C at 6 o'clock news. It took the Germans by surprise. Immediately they captured the Italian G.H.Q outside Rome, took 30 Generals as prisoners. In the north, the Germans had as per their contingency plan arranged for four Divisions to hold the passes under F. M. Rommel.

In the south, where the VIII Army had landed, there was an amusing scene when an Italian General claimed seniority and wished to take over command of the Allied army. Monty was not amused. Not much help was expected from the Italians nor was it forthcoming. In practical terms, General Chand Narain Das tells me, the Italian Partisans were helpful and due to their contacts with the local populace many, Indian POWs, held in Italy, got a chance to escape. The result of the Italians opting out, was that we had to fight the highest grade of German troops, their Airborne and SS who prided themselves as the upholders of the motto "Loyalty is my Honour," and were equipped with the most sophisticated weapons.

Italy, the land of sunshine, as described in travel brochures being mountainous with numerous rivers, fighting in it was an engineer's nightmare. The roads in Southern Italy cross numerous rivers—twist and turn—in the mountainous country. The roads and the railways were destroyed by the retreating Germans with great thoroughness. They used "Rail Rooter," a machine specially designed for ploughing up the sleepers and demolished bridges to hold up our advance.

A sea borne attack by the US Army further up the coast at Salerno having been held up, the campaign had to be fought patiently up along the spine of Italy. It was an engineer's war. Besides the normal complement of divisional engineers, a large number of Engineer Corps Troops including two Indian Engineer Battalions took part in this campaign. A total of two thousand four hundred and ninety four (2,494) Bailey bridges were built.

Whilst the British VIII Army, after landing at the port of Taranto, advanced along the Adriatic Coast, the Anglo-American V Army which landed at Salerno, on September 8, and occupied a beachhead by early hours of the next day, was held up by the Germans. Jodhpur Light Infantry Battalion helped the U.S forces at the beaches. The Indian Army's first DSO of this campaign was won by Major Ram Singh of Jodhpur Infantry.

The first Indian formation to land, the 8 Indian Division, was commanded by Russell Pasha[5] ex 6/13 Frontier Force, whose only wish was to command this division. It was an unopposed landing. At Taranto on September 19, they landed to the sound of blaring music from the Italian Naval Ships in the harbour. The Division became part of the V Corps of the VIII Army of which the other Division was the 78 British Division. They advanced along the Adriatic coast. The American V Army having landed at Salerno, met with the VIII Army and advanced along the West coast towards Naples.

The weather changed, instead of a fine warm and sunny Italy, it became wet and cold with icy rain. The innumerable rivers rose in spate. At 0345 hours on November 29, the 6/13 Frontier Force Rifles attacked. They had to wade through chest deep icy cold Trigno River. The Division took three days to go through it. Then faced another formidable obstacle at Sangro River. On the escarpment on its far side were five villages, their stone houses surrounded with wire, had been heavily mined—made a strong redoubt. Two of these hedgehogs, villages of Mozzagoranga and San Mario were in front of the 8 Division. These were well covered with fire and mined heavily with Tellermines and Schumines of which the latter were deadly as they jumped up three feet above the ground and then burst killing relentlessly. The position was held by the crack German 1st Parachute Division.

The bridging of the river in flood was a terribly formidable task. But tanks had to get across to deal with the flame throwing enemy tanks. With gigantic effort, the crossing over the Sangro River was achieved. On November 21, 7 Field Company of the Bengal Sappers—built a Bailey bridge across this torrent. The Field Park Company lifted 261 Tellermines in two days. On the night of November 28, the 8 Indian Division captured Mozzagoranga in a swift operation. On its left the New Zealand Division secured the crossing by first light on the November 29. In the early hours of the November 30, 1/12 Frontier Force Regiment attacked with the support of tanks. The enemy held strongly defended stone houses. These were tackled by first firing A.P. (Armour Piercing) shells to make a hole in the walls; followed by a H.E. (High Explosive) shells with a delayed action fuse to detonate inside the houses. The German winter defence line was thus breached by Super human efforts.

The advance continued till the crossing of the next river Moro. This river, running in a deep ravine, posed a formidable problem. The destroyed bridge on it was located at a right-angle bend in the road. There was no space on our side of the bank, to launch a bridge from the home bank. With great ingenuity, this very difficult task was tackled by 69 Field Company of the Bengal Sappers. It was launched from the enemy side, while the 3/15 Punjab and a company of the Marathas provided cover to the Sappers. The building of *"The Impossible Bridge" was so appreciated by the Commander—that he presented the 'Pennant' of his staff car to Company Subedar of 69 Field Company.*

The next obstacle on the way was Villa Grande. The Essex Regiment had tried but failed. The 3/8 Punjab under the guidance of

Major Girdhari Singh used an unusual stratagem to dislodge the Germans. Using some captured Spandus, Girdhari arranged to fire incendiary bullets at a couple of haystacks in the village. The Punjabis then advanced under the cover of the dense smoke and captured the position.

At the end of the year, General Montgomery went off to command the invasion forces in the West. At his farewell address in his staccato speech he claimed, "My VIII Army's tails are right up." At this, Johnson, the BBC correspondent, in his rich Dublin brogue interrupted—*"Excuse me General, up what?"*

The 8 Indian Division settled down for the winter in the mountains till it was called up for the final attack on the Gustav Line, based on the Cassino massif.

The Fourth Indian Division at Cassino

The Monte Cassino story is hogged by two teenagers—a Lieutenant Iqbal Krishna of the Rajputana Rifles who told me how it fell. To lead an attack on the famous Monte Cassino—the inferno which consumed some 4,000 of our finest soldiers of 4 Indian Division and Sepoy Kamal Ram, a fearless Jat from Haryana, by his gallant foray opened up a breach in the enemy defences which enabled the 7 Indian Division to cross the Rapido river and break the German defenses. Unscratched in battle he lived to wear a Victoria Cross.

The 4 Indian Division came up after six months rest in Palestine. Captain Hari Shankar Kapur of the Piffers told how during this time he was detailed as a liaison officer to the various Princes visiting their troops. His first assignment was with the Maharaja of Sirmoor State. As the conducting officer he stayed at the Shepherds at Cairo. It was a very pleasant change. Later, he looked after the Maharaja of Nabha on his trip to Haifa and Jerusalem—they stayed at King David Hotel. The most interesting and colourful was his assignment with the Ruler of Jaipur. They were entertained lavishly by the famous jeweller Ganeshi Lal, at Cairo.

When Hari Kapur's FFR Battalion was a part of the South African 6 Armoured Division allotted to the American V Army of General Mark Clark, one of the cooks of the Battalion was accused of raping an Italian girl. For the Court Martial held at Florence, Hari the defending officer of the cook and Qazi, the defending officer of the girl, the second in command of the Battalion; all went down to Florence and checked in at the local guest house of the Armoured Division, situated on the bank of the river.

The three of them were having lunch, when they were called to the reception desk and told that whereas the Britisher could stay, the two Indians would have to move out because of the 'Colour bar'. Qazi and Hari had to move to the YMCA. A bad case of Apartheid, their senior officers later apologized for the shoddy behaviour of the staff.

Cassino—The Key To Gustav Defence Line

The advance of the VIII Army was held up at the Gustav Line, blocking the way to Rome. Twenty miles inland from the Adriatic, at the head of the Valley stood a great stony mass—Monte Cassino. On it was the famous monastery, which was the lynchpin of this position. An excellent example of control exercised by high ground in a battle zone. This ancient and magnificent monastery was strongly held by the Germans. In front of it flowed the Rapido River. The main road to Rome went through the valley—dominated by fire from the monastery held by the Germans. This highway Number 6, from Naples to Rome, could not be used as long as the monastery was in German hands—they could almost toss a stone down onto this main arterial road.

The Americans had tried to dislodge the Germans, having failed to take this stronghold, our 4 Indian Division and part of New Zealand Corps was given the task.

Major General Rees, ex-Rajputana Rifles—the Divisional Commander, in his spare moments had bought a book on this monastery at one of the old shops. On studying it, he was quite convinced that the position with its massive 12-feet thick walls could not be captured without a tremendous amount of softening up. The proposal for its bombardment was however not acceptable to General Mark Clark of US Army, the Army Commander. On the morning of February 15, under a clear sky, with a token opposition from the Germans, an armada of bombers, made their appearance over the monastery and for four hours wave after wave of bombs pulverised the building. The planes dropped their bombs from heights. It was an impressive and awe-inspiring spectacle. To add to the damage, our Artillery also started to batter away at the ruins. The Germans did not yield and used the ruins for putting up defence.

The Youngest Survivor

Iqbal Krishna (my nephew) had joined the Indian Military Academy just before the start of the Second World War. At its outbreak, the training was curtailed and the cadets were commissioned early. He, thus, found himself as an officer at a tender age of sixteen and a half. They were referred as *bacchas* in the Battalion. With the shortage of officers in the Infantry, he was soon on his way, to the Middle East.

After a year in the famous 4/6 Rajputana Rifles, part of 5 Brigade of the 4 Indian Division in the Desert, the battalion moved to Italy. A Divisional attack was planned from the rear of the American positions, west of the monastery, to coincide with a frontal attack by the New Zealanders from the south of the town. Iqbal was commanding a platoon in "C" Company of Jats. In this junior position, he could only recall the long march, the approach to battle through the town of Cassino. It was a cold and eerie experience as most of the buildings had been reduced to rubble. Throughout their move up, they were subjected to intermittent fire and suffered several casualties. The worst affected were the mules, out of about 200 mules, only 20 reached the forward position. Mules are an abomination in battle—they are awkward and cannot be guided to take cover. The attack took place on the night of the second day.

Recalling the final assault, Iqbal told me that although from a distance the mountain appeared bare and smooth, it was a jumble of hollows and crests. The enemy had set up Machine Gun posts in the ruins, and swept the slopes with fire. Our casualties were thus very severe.

The timings were kept up, B and C Companies of Jats went in at 2350 hours. They scrambled forward moving from bound to bound, making good progress and reached a position where he could discern the outlines of the monastery. Then suddenly, out of nowhere, arose a German at a range of hardly ten yards and pumped in a burst of 9mm rounds, into him as he was leading his men. That was enough to instantaneously knock him down senseless. No doubt his Jemadar, Manik Ram rushed with fixed bayonet killed the assailant. The attack went on, they almost got to the monastery but could not hold on to it against enemy counter-attack.

After being hit, Iqbal does not remember a thing, about how he was rescued from the battlefield. He found himself in an Advanced Hospital established in another monastery. He was showered with abundant care, everything including a brand new shaving kit had been laid out alongside. Luckily, the bullets had gone right through his left shoulder and there was no need for surgery.

A fortnight later he was taken by air to Base Hospital at Cairo. While convalescing at Cairo, he struck a friendship with an old British Colonel, who was working at General Head Quarters, Cairo, from whom he got his first lessons in staff duties.

Attempts at the capture of Monte Cassino by the 4 Indian Division went on right through the winter. A second attempt was made on February 19, but the weather did not help. The last attempt code named "Brahman" was on March 15, by the 7 Indian Brigade. For this a special effort had been made by the Sappers to make an approach track for tanks to be able to support the attack. This track almost a mile long, was constructed under the able supervision of Colonel D. R. Stenhouse, the CRE 4 Indian Division, my course mate at the "Shop." It was called the 'Cavendish Road' after the then Commandant of the Bombay Sappers. The attack was delayed as air support was not possible in the thick fog. Eventually, the powers to be, decided that the 4 Indian Division was exhausted and needed rest—it had lost 4,000 men on the Cassino alone without significant gain but defences had been considerably weakend.

The clearance of the mines on the tracks led to a most gallant action by a Subedar Subramanyam of 11 Field Park Company of the Madras Sappers. While he was operating the mine detector a Lance Nayak accidentally trod on a Schumine. Before the mine could explode, the Subedar flung himself on it to save his colleagues. For his great act of sacrifice uncaring of own life, he was awarded the George Cross.

The Gustav Line

Finally to penetrate the Gustav Line a strong Polish Corps, the 8 Indian Division and the 2 Canadian Armoured brigade were secretly transferred from the Adriatic side to have a crack at this tough nut. The position was overwhelmed by a general offensive in May. This time there would be no frontal attacks. The Polish Corps struck into the hills, while the 8 Indian Division along with the Canadian tanks practiced river crossings on the Volturno River, 20 miles south of the river Gari—using new Mk III assault boats. (River Rapido is known as Garigiliano –Gari for short in the lower reaches)

On May 1, the 8 Division and the Canadians moved to the left bank of the river—on the left of 4 British Division. The right of the Division was only about two miles from Cassino. The Division then lay dug for twelve days while the weather improved and the land dried out for the tanks to operate. Sappers including 7 Field Company of the Bengal Sappers meanwhile lifted the mines on this side of the river and built hard roads across the soggy fields. The men rested during the day and worked on their tasks at nights. The battle started on the night of the May 11/12, as the moon rose at 11 p.m. Eight hundred guns, opened up and pounded the enemy positions. The 4 British Division and 8 Indian Division with the Canadians crossed the river Gari.

The southern part of the Gustav line was entrusted to the 8 Indian Division and the Canadian Armoured Brigade in Sherman tanks. The crossing of this typical fast flowing river was very strongly defended by the German Para Herman Goering Division, who had sited their Spandaus to fire just above the water line. The crossing was accomplished at night, with 17

Brigade on the right with 1/12 F.F and H K Sibal's 1/10 Baluch.[6] Two of the officers of this Battalion Himmat Singh and Gnarre Singh played a conspicuous part.

19 Brigade on the left, with Kashmir Singh Katoch's[7] 6/13 and 3/8 Punjab on the right, the attack was preceded by a heavy creeping barrage. Kashmir Singh Katoch was commanding the Dogra Company of the 6/13 FFR. They were taken across the river in assault boats, manned by the Sappers. The Pathan Company suffered a number of casualties, but his Dogras got away fairly lightly. They were held up by a machine-gun nest. Jemadars Dhanak Singh and Bhan Singh, the first two Platoon Commanders, were killed in the attempt to dislodge it, the third Hoshiar Singh was lucky as Katoch and he between them pressed home the assault and blotted it out. Katoch was awarded a well-deserved Military Cross.

The 7 Punjab had trouble with their boats, fighting on their front was very fierce. Major Sujan Singh's D Company and a whole of a Sikh platoon were wiped out. B and C Companies under Major Girdhari Singh were held up by machine fire, when Kamal Ram, a 19-year-old Jat, crawled through the wire, successfully shot the gunner, bayoneted the feeder, swung round to shoot an officer who came at him from behind. He, then went on to deal with four Machine Gun posts and came out unscathed, as the youngest winner of the Victoria Cross.

The Germans were forced to retreat. A special technique tried during these operations was the use of ARK bridges. "ARK" bridges were built using American tread way tracks on Churchill tanks. Single ARKs, with tread ways at both ends were positioned to take Sherman tanks across the river, in minutes despite intense enemy fire.

One of the bridges called the Plymouth Bridge was a triumph of improvisation, a span of Bailey Bridge was mounted on a tank—the tank was driven into the river to support the bridge.

The Polish Corps attacked Cassino. The town was cleared on May 18 when they took the position and hoisted their flag on the ruins of the monastery. The monument in the Polish Cemetery to the 3,784 braves killed, poignantly reads:-

We Polish soldiers
For our freedom and yours
Have given our souls to God
Our bodies to the soil of Italy
And our hearts to Poland

The Americans were very keen to claim the capture of Rome. It fell on June 4 to U.S V Army. Once in Allied hands, regular parties of our men, Hindus, Sikhs and Muslims—could be seen being conducted round the Vatican—guided by Hindustani speaking Italian priests. One of the rewards for outstanding service was a holiday in Rome. It had been declared an open city—so the bridges were intact.

The Gothic Line

This defensive line in the mountains between Pesaro on the Adriatic and Spezia on the Gulf of Genova was the last hurdle. It barred the way to the Valley of the river Po.

At the end of July 1944, King George VI paid a visit, and saw the Indians at battle of Anezzo, He personally pinned the VC on Kamal Ram's chest.

The 8 Indian Division reached Florence in mid August 1944, the city was declared an open city to preserve its art treasures. The advance to the Gothic Line started in early September. The advance was now slower as seven divisions had been diverted for Normandy Landings in France.

The VIII Army, including, 4 Indian Division, moved back to the Adriatic Coast. General Chand Narain Das, who was commanding a Company of machine-gun Battalion of the Rajputana Rifles at Ortono, told me of some interesting escapades. One was a very near miss—due to the sagacity of his Jeep driver Hardeva. They were going on the road, when, a shell landed in front of them, soon after the second landed just behind. Hardeva immediately realised that they had been bracketed—and drove off the road at the fastest possible speed. Sure enough when they looked back the next shells had landed where the Jeep would have been—if they hadn't got off.

Through the winter, the Germans kept them under intermittent fire. They were lucky living in a low building, the Brigade HQ in a taller building on the other side of the square received more than its share of shells. One day, Chand received a message that his Havildar Major had been wounded so he took him to the Advanced Dressing Station (ADS) manned by none other than our neighbour, Ved Parkash. Unfortunately, he had been hit by a shell splinter, and could not be saved. It took time to arrange for his cremation. At last when Chand got back and lay down in his bed, his head hit a couple of sharpnel pieces on the pillow—a lucky escape as these had arrived during his absence to the A.D.S. The Germans, it was discovered, were trying to expend their stocks of shells before retreating. They were followed up on October 19. There after 4 Division moved to Greece, leaving the 7 and the 10 Divisions plus the 43 Gurkha Lorried Brigade to finish the job.

The 8 Indian Division having advanced 240 miles in five weeks was relieved by 10 Division and went to a rest area, for the winter. They billeted in the mountains north east of Arezzo. One of its Brigade, the 6/13 Frontier Force whilst at Monte Cerero was surrounded by the Germans. and had some anxious moments. Recently, Mrs. Satarawala—wife of late Jhangu Satarawala, of the regiment and now living at Pune received two photographs from her husband's colleagues from the Battalion. One showed a monument built by the village folks of Monte Cerero to commemorate the gallant defense of the area by our troops. Another was of an incident during their encirclement. When they wanted to send a message to their Headquarters, they tied it to the leg of a carrier pigeon. The pigeon instead of taking off—perched on the battlements. Jhangu in contravention of the orders fired his revolver at the bird. It at last took off, delivered the message—which brought them the necessary help.

There were also some lighter moments. The Sappers during this time were developing tracks forward near Valenzano. During the building of a track Jemadar Jaswant Singh of 5 Field Company, captured four Germans as prisoners from a house overlooking the track. The platoon officer was not far behind—he rescued a Countess from a Chateau nearby, and the next day at her request also brought back her maid much to her delight and received her favours.

In July 1944, the US brought in additional troops, to compensate for the divisions diverted for "D" day. These included the 92 All Negro Division with white officers, and 25,00 strong Brazilian Brigade. These troops with no combat experience became a liability. In December, the 8 Indian Division moved behind the 92 Division as backstop to bolster the young troops when they got attacked on December 26[8].

The Tenth Indian Division

In the autumn of 1944, 10 Division followed up the withdrawing Germans. The main effort against the Gothic Line was by this division and the newly arrived 43 Gurkha Lorried Brigade alongside with the British 1st Armoured Division. Pesaro on the right flank of the line was turned, the VIII Army entered the plains just short of the Senio River. By November it was at Marradi. The countryside was flooded by the enemy to block the advance to the Po Valley. The GOC 10 Division, wished to have communications with the troops on the far side of the river, from which the enemy had not yet been entirely driven out. To meet this requirement, "Houdini" bridges were invented. These consisted of a pair of steel cables taken across the river, through a couple of snatch blocks, and back-to-back tackles on the near bank. The decking was made of timber, salvaged from the houses and snake-lashed to the cables.

These bridges could be submerged by day, and used only during the night. On the shore most decking panels were removed at daylight; leaving only the cables, which could be camouflaged with the thick brushwood, which grew on the banks.

For footbridges, I recall someone telling me that they used the aluminum electrical transmission cables, which were available from non-operational power lines!

The Senio—1945—two more VC's—of 8 Indian Division

The final breakthrough: the spring offensive of the VIII Army opened on the afternoon of April 9, 1945. It started with a massive bombardment followed by a mighty Artillery barrage on the defenses of River Senio. A special feature of this operation was the six minutes of intense flame throwing—just before crossing the start line. The 8 Indian Division under General Russel (Pasha) attacked across the River Senio on the same evening soon after seven, led by tanks and flame throwing "Crocodiles." The 1/5 Marathas of 19 Brigade were on the left and 6/13 Frontier Force Rifles of 21 Brigade on their right. The forward companies of both the battalions were held up by very heavy fire. The fire was so intense that assault boats, could not be used, so the troops waded across. The advance was made possible by the bravery of two Jawans, both won VC's; the 20-year old Namdeo Jadhao of the 1/5 Marathas, the Company runner was the sole survivor, he dragged his two wounded comrades to safety, dashed forward, silenced the German crew of the spandau with Tommy gun fire and took two more posts with grenades, climbed to the top of the bunker and shouted *"Shivaji Maharaj Ki Jai."* The reserve company then crossed and occupied the west bank.

On his front, Ali Haider of the 6/13 F.F. Rifles attacked single handed, with grenades and Tommy gun, and destroyed German strong points holding up the advance. The heroic deeds of these two men form a glorious chapter in the history of the Indian Army. The 43 Lorried Brigade poured through the hole punched. The 8 Indian Division secured a bridgehead over the River Po without much delay on 26 April . The fast flowing river was over 200 yards wide. The Sappers were given a new type of raft with four outboard motors at each corner. These were found to be difficult to manage at first—but our men of 7 Field Company Bengal Sappers soon caught up. It was spanned by a 1,370-foot pontoon bridge—the longest bridge in the Italian campaign.

Divisional engineers were the first to reach the next river, Adige, on 28 April . They built a 490 feet Folding Boat Equipment (F.B.E) bridge—which was named Roorkee Bridge— and opened at 9 am on April 29.

No account of this lightening move forward would be complete without mentioning the role of the Reconnaissance Regiment—the 6 Lancers. Towards the end of the Italian Campaign, after crossing the River Po, the divisional Cavalry Regiment the 6 Lancers, fanned out towards Rovigo. There were a few German stragglers on the way, who were rounded up and sent back to the Headquarters.

Soon they were near Canale Bianca, a small twenty-yard wide obstacle that could have held up the advance. The Commanding Officer received a message from Jemadar Fazal Dad that he had captured a wooden bridge across it. The enemy had tried to burn it, but the fire was put out and timbers replaced. This bridge called "Fazal Dad" bridge took the whole of the Division across. Thereafter it was all kisses for the handsome, wine and roses. The Germans gave up on May 2, when all German troops under F. M. Kesselring surrendered. It was a massive victory crossing the four major rivers all within three weeks—22 days to be precise. The war was over six days later. A large number of decorations were awarded. Some notable ones were: MC's to Iven Surita and Mian Dad, George Crosses to Subedar Subramanyam and Sowar Dittoo Ram. The 10 Indian Division Engineers alone won thirteen MC's. Lieutenant Colonel Mahadeo Singh won a DSO.

Notes

1. Later General
2. This operation has also been dramatised by Irwin in his book *The Desert Fox*
3. Major General Khurana I.M.S
4. *The Tanks* by Capt Liddel Hart
5. Lt Gen Sir Dudley Russell
6. Lt Gen H.K. Sibal was then D/Q of 17 Bde.
7. Lt Gen Kashmir Singh Katoch M.C
8. *Circles of Hell War in Italy* by Eric Morris

❮5❯

With Chinese and Americans

There have been many ups and downs in our relations with China. Many only know of and are wrongly obsessed by our reverse in 1962. Few know of the last century, when our Army went to China as conquerors, and only fifty years ago we fought alongside them in Burma. This is about the Second World War when our military co-operation was at its best.

Even in 1962, there were some lighter moments. Gen. Manmohan Khanna told me of his Kumaoni Company runner. It was dusk, he jumped into a slit trench for a breather. He felt like having a smoke, but found he had run out of stock so he whispered to the one in the next trench *"Sathi Beedi Hai"*. The fellow occupant replied, *"Hindi Chini Bhai Bhai."* He quickly got hold of his Sten and jumped out and away.

For our friends the Chinese, the Second World War had started in 1936 when Japan invaded the mainland.

Their first action was the great sacrifice made by three Japanese Sappers who charged in with a live 'Bangalore Torpedo' type of device to make a breach in the barbed wire at Shanghai. Their great deed is even today remembered in a Shinto shrine.

The Americans with a world wide prerspective realised the tremendous effort being made by China to stand up to Japanese attacks. The Chinese Republic established after the revolution in 1912 did not get a chance due to internal troubles. Even in the early thirties when we as cadets, were under training along with our Chinese colleagues, Seng Kung and at the Royal Military

Academy at Woolwich, China was fighting bandits or warlords. Yet, China because of its size and population was the biggest bastion against Japanese expansion.

It was therefore but logical that one of the war aims of the Japanese entry into the Second World War was to interrupt the overland supply line to Chunking (China's war time capital) through Rangoon and Lashio by rail, and then by road to China.

When we were very hard pressed in Burma and were being pushed out of the country, short of troops, we were unable to hold the Japanese. Marshall Chiang Kai Shek, and his wife, paid a visit to Delhi and arranged to send two of his Armies (Chinese Army's were equivalent to our Divisions) to fight alongside us to hold the Japanese south of the Prome-Tungoo line. General Slim, who had arrived in the theatre, was holding Prome, with the 17 Indian Division. The Chinese moved in their 36 Division to hold Tungoo on its east. This place was of particular importance as it overlooked the vital bridge over the river Sittang, an important point, on the railway line from Rangoon to Lashio. The main route for supply of American Lend Lease materials to China.

The Chinese armies came in great force from the Yunnan, across the river Salween to Lashio, they were received with great joy by their compatriots in Burma and were temporarily housed in very tastefully constructed Bashas–bamboo huts. A most interesting feature of their force was the inclusion of girls, who worked as hard as the men and carried full loads. Also present were some boys in yellow uniforms, who acted as orderlies and ran errands. They were also known to be capable of using weapons at times. One of their Commanders' had been educated at St. Cyr in France.

The Chinese were placed under command of an extraordinary Lt General Joe Stilwell of the United States Army. He had a difficult task as he wore two hats—he was also the adviser to General Chiang Kaishek and had thus to get his approval.

A conference was held on April 9, 1940, at Maymyo on the Mandalay-Lashio road at which both the General–Issomo Chiang Kaishek, General Stilwell popularly known as "Vinegar Joe" and British Chief were present. Stilwell was however under the overall command of Gen Slim", the Army Commander. In a conference he told Slim "I have been a good subordinate having obeyed all your orders". There upon Slim had retorted—"yes, after ensuring that I ask you to do what you wish to do'. The Japanese bombed the location but not much damage was caused.

The Chinese 36 Divison occupied the defensive position at Tungoo; the Burma Division by then, low in strength, was withdrawn behind the 17 Indian Divison for rest. The Japanese reacted very strongly and launched very heavy attacks against the Chinese. Another Division the 22nd was readied at Lashio, to come to its help.

The fighting was so severe that the Chinese suffered their greatest number of casualties whilst at Tungoo but after about a week, they had to retire. They could not hold as the Japanese having refloated their craft in the Rangoon harbour, were able to bring in tanks against which they had no weapons. Further two Japanese attacks on March 21 and 22, on our planes whilst still on the ground at MAGWE, had crippled the R.A.F. Our troops, denied air support, could do little in the face of overwhelming opposition except to retreat to fight another day.

The Hukong Valley

General Stilwell at the age of 60 walked back all the way out to India through the gruesome Hukong Valley. A 135 miles of a green valley, through which came back a large number of our Army personnel including my brother, civilians and the two divisions the 34 and 22 of the Chinese Army. They suffered great hardships—shortage of food and water—attacked by leeches, and mosquitoes, they all needed urgent medical care on arrival.

The two Chinese Divisions were our 'guests' at Ramgarh Camp where they were rested and refitted by the United States Army. They returned to open the Ledo road.

The Ledo Road

On the loss of the sea route, the supply line was maintained by the United States Airforce, across Assam (over the HUMP), it was slender and expensive.

The establishment of the road to China, across Burma, was therefore of utmost importance to America in terms of the overall war effort. A tremendous effort had been made by the Indian Government to build 200 airfields in the North East of India.

The most known of these were Chabua, Tulihall, which enabled the USAF to maintain the supplies over the "HUMP".

Working against great odds, Stilwell Army had established themselves at Shiagbwiyang in the Hukong Valley, 103 miles away from their base at Ledo by December 23, 1943. They had by November 15 surveyed up to MS 99, the dozers were at MS 79 and the metaling had been completed till MS 48 when Brigadier Lewis Pick (an appropriate name) took over the construction.

Progress was quicker thereafter as they caught up with an old trace of a road built by the Japanese.

Another force, which is sometimes overlooked, is "Merrills Marauders" of the United States who came into the arena on March 12, 1944. The official name of this force of 3,000 Commandos was American 5307th Composite Unit. Their first task was to attack the Japanese defences at Jambu Bum at the southern end of the Hukong Valley.

Eventually, they moved forward and captured the airfield at Myitkyina by May 17, 1944. The town was held very tenaciously by a Japanese garrison of about 1,500. They held it for 78 days. It was essential to dislodge them. Recourse was made to ask for 200 volunteers from the Chinese on the promise of rupees thirty each and double promotion. It worked except that the Japanese were allowed to escape through the get away gaps.

With the capture of Myitkyina progress was better along an old trace. The next hurdle was the rail town at Mogaung—which in spite of all the problems of sickness was captured by Mike Calvert of the Chindits, with the help of his Gurkhas, two of them won the Victoria Cross. After this last effort, the Chindits were withdrawn and 36 British Division was landed in Burma to travel along the railway corridor down from Myitkyina towards

Mandalay. This move prevented the Japanese from interfering with the construction of the lifeline to China.

This Division consisted of two British and one Indian Brigades. They were supported by two Field Companies of the Bengal Sappers and an Engineer Battalion. It was provided massive air support by Col. Cockraine's First Air Commando with a total of over 400 aircraft—including 30 Fighter-Bombers, 12 bombers, 25 transport planes and some 100 light aircraft.

Shweli Crossing

This is the story of the Ledo Road and the part played by the 36 British Division which helped in opening the road to China by operating in North Burma along the railway corridor from Mytchina, pronounced as 'Ma-cheena', the northern terminous of Burma railways, towards Mandlay. During these operations it was necessary for the Division to cross the river Shweli. This river crossing is of interest as we have the eyewitness accounts of two of the young subalterns who took part in it.

The 36 Division of two British and one Indian Brigades was under the command of General Festing and came under the control of American Northern Combat Area Command, headed by Lt. General 'Vinegar Joe'. It was supported by two Field Companies and an Engineer battalion of Indian Sappers. This division had been flown in and was dependent on the lavish air support from U.S First Air Commandos. It advanced down the railway corridor south from Mytkyina, west of Irrawady.

Lieut. Rau[1] of 30 Field Coy Bombay Sappers was the first Indian officer in this Division, till he was joined by Bhaiya Rajwade[2] of 58 Field Coy Madras Sappers.

The Division after a peaceful crossing of the Irrawaddy was held up at the crossing of the Shweli. The story of this crossing, in Rau's words, amply shows how well our officers met the challenge. "General Festing, the Divisional Commander, had expected light opposition." The two Japanese Regiments (equal to our Brigades) put up stubborn resistance.

Their concealment and camouflage was so good that we were unable to locate them. The first to cross over was the British battalion–the Buckingham Shire Light Infantry—generally called "Buffs". They started crossing in broad daylight, supported by a Sapper platoon from 58 Field Company, who rowed them across in assault boats. Lieut Bhaiya Rajwade was in charge of the Ranger boats, which were rather unwieldy and slow. The Japanese let the first two Companies get across, but when the third Company was crossing, they opened heavy enfilade Machine Gun fire from an island in the river. The crossing was soon in shambles, our Thambi's gallantly rescued the British soldiers, who were wounded and drowning. The two Companies which had crossed, had hardly dug in, when they came under repeated "Banzai" attacks. It soon became apparent that they could not hold on there. They were recalled, under cover of mist and a smoke screen, a few swam back. Rajwade remained on the far bank, through the day and night, most of the officers of the Buffs had been killed or wounded. He gallantly organised the defences and evacuated the wounded. He was one of the last to swim across the wide river, and was awarded an immediate Military Cross. Four of his Sappers won Military Medals.

The Divisional Commander then decided to cross further up the river, with 30 (Rau's) Field Company to support the crossing.

Muddy Rau recounts-"The river here ran three miles on a straight course. Unknown to us, there was a Japanese Observation Post (O.P), on top of a tree, from which was seen a grandstand view of this stretch of the river. The crossing started smoothly—the Royal Sussex Battalion crossed under the cover of darkness. At daybreak, the Japanese F.O.O (Forward Observation Officer) at O.P became active. Every time a raft would be pushed off the bank, Artillery fire was brought on it, spouts of water would rise all around it. Strangely enough, there was only one direct hit. The river was about 400 yards wide; each raft attracted half a dozen shells. The Officer Commanding insisted that every raft must have an officer on it, and personally took turn. Each of us did over a dozen trips; the only casualty was one wounded by a shrapnel. After mid-day they started shelling the jetties, which they should have done in the first instance. In spite of them being obscured from direct vision, our casualties mounted. We lost a V.C.O. In the evening I was sent with two sappers to sort out the cluttering up of the far bank, by burntout vehicles, and dead mules.

A battle developed, the Japanese launched a counter-attack. It was getting dark in the confusion; I got separated from my men, who as I learnt later, swam back. I jumped into a trench, occupied by two British soldiers. I was not exactly welcome, but they soon became friends with typical cockney humour.

The Japanese were all around, and were putting in their attacks. I was nowhere near the point of attack, but fired off my revolver defiantly, to boost up my morale. There was nothing to eat but bully beef, after surviving for sometime on water, I gobbled it up. The two Tommies solemnly shook my hand, at my initiation.

They kept on grumbling and joking alternately. One was henpecked and kept on talking of his battleaxe-wife back home, whom he loved dearly. Next evening, I was called up to the Battalion Command post where the C.O. put me on the radio to my Officer Commanding. First he cursed me for not having come back to the unit. I could not tell him how scared I was of swimming back. He told me that he would be sending four Folding Boats next night and I was to arrange to get back as many wounded as possible. I was keyed up to an edge. The approach to the river involved crossing a sandbank about two hundred yards wide, dragging stretchers across would be difficult. The Battalion Commander sent out guides with shielded torches, to direct the boats into some reeds on the bank. There was also the problem of the wounded making noise; and we did not have enough shots of morphia. Someone had the bright idea of getting the not too seriously wounded, dead drunk into a stupor—there was plenty of rum, it worked to quieten them.

At about 2300 hrs, Havildar Rab Nawaz Khan, reported to me, his arrival with four boats and crews. Due to shortage of stretcher bearers we made four batches. I went with the first batch, the wounded were loaded, I was determined to keep the four boats together, this involved certain delay. One of the wounded BOR'S called me a callous bloody Indian, but I swallowed it.

When the fourth and the last batch was being dragged across, the Japanese fired a parachute flare which lit up the whole area, and started firing. Luckily, the sandbank had some burntout vehicles and dead mules, which provided cover. I, at the tail end of the batch, crawling, close to the ground, felt that my web equipment was hindering me so I took out the revolver and left the belt and the holster. The din was terrific, the enemy fired Very lights, and flares, but shadows play tricks, they never spotted us. For all their trouble, they only wounded one stretcher-bearer, who was also evacuated in one of the boats.

The Japanese, thereafter concentrated their fire on the far bank jetties. It was clear that we could not land there. The boats were fully laden, with only about six inches of free board. All able-bodied men including myself, got into the water, and swam alongside. The boats were allowed to drift. It was dark; we were looking for a suitable landing site. We went down about two miles. Dawn started breaking; to avoid Japanese patrols, we banked. There were cliffs at the place where we finally landed with great difficulty; we carried the stretchers to the safety of the jungle. The effect of the morphia was wearing out, there was quite a noise, interestingly, the least wounded, were loudest. The R.A.M.C. (Royal Army Medical Corps) Major, incharge of the ambulances was on the look out, we contacted him and were relieved, when the doctors and their nursing orderlies took over. He shook my hand, and said I had saved quite a few lives-I felt a sudden glow of pride — on a job well done."

The 5 Engineer Battalion, using timber directly from the forest and their hand tools constructed a 1500 feet, teakwood trestle bridge across the fast flowing river, with a bed of soft shifting sand. It took them only eight days. A magnificent job, it solved the maintenance problems of the Division once for all.

It's a *matter of great regret that there are no bards to sing. Their praises today, we no longer have such units.*

Mine Clearing

After the Shweli river crossing, we resumed advance south-east from Myitson to Mongmit, in the first week of March 1945. My platoon was given the task of clearing mines and other obstacles from the road.

As we were approaching Mongmit, we entered a Burmese village. It consisted of the usual wooden huts built on stilts, in fairly good shape. As we were entering the village, a local came running up and in an excited voice told us (we had an interpreter with us) that the Japanese had buried some boxes in the road and that the villagers had stuck sticks as markers near the spot where the Japanese were digging. There was no love lost between the local Burmese and the Indian and British troops, they wanted the suspected area cleared. We soon came across one of these sticks stuck in the road. My platoon, Havildar Rab Nawaz Khan, had been awarded the IDSM in the Middle East where he had cleared a number of German mines and deloused numerous booby-traps. He felt that if he could tackle the far more sophisticated German mines and booby-traps, the crude Japanese versions were child's play to him. It was about 1400 hrs he and I gingerly walked upto the stick with a long signal cable in our hands. We scraped the earth carefully away and this disclosed a wooden piece about 12 inches square with another wooden piece of similar size below, separated by 2 bamboo pieces. We carefully tied the cable round this contraption and went back about 50 yards behind a ridge in the ground, which would give us ample cover, pulled the cable nothing happened. After waiting a few minutes, we went back found the top piece lying some distance away and two wires coming out of the ground. The Havildar then asked me to get behind cover while he investigated the mine further. I stayed, I suppose I wanted to show my men that I was not scared. The Havildar eased the earth away which disclosed a wooden box with two wires coming out of it. The box obviously contained explosives. He made the hole much larger so that there was sufficient room to get the hands to the bottom of the box.

He, then, brought out a small piece of mirror, which he carried, passed it underneath the box at an angle to see what was underneath it. He could see nothing but earth. The mirror helped but was not foolproof. The tension was tremendous, we were both sweating profusely by this time, scared of a release type booby-trap, which would go off when the box was lifted. Finally, all the earth was removed from under the box. It had taken nearly an hour and we gently lifted the box out. It weighed nearly 20 lbs. Luckily it was not booby-trapped.

Just at this time, my Company Commander David Wilson came up on his jeep and I proudly handed the box over to him. He realised what it was and gave me a tongue lashing which I still remember. He straightaway got a pair of pliers and after cutting the wires short, insulated them. Explained to me that if the two naked wires had made contact, I would have blown to smithereens. We carefully dismantled the box and found it contained a couple of dry cells which were connected to an electric detonator embedded in the explosive (Picric Acid).

"There were two planks with contact points connected to the dry cell separated by two bamboo pieces which would crush and activate the mine on load being applied. A simple design. Later on, I had hardly walked about 40 yards when there was a terrific explosion. My helmet flew off and I was stunned for a couple of minutes. I would like to mention here that we never put down our chin straps since old timers warned us that we may be strangled by the blast when the helmet lifted off. As soon as I regained my wits, I turned round to find out what had happened to my Havildar. For the next hour, my men were collecting bits and pieces of the gallant man".

36 Div was soon able to contact 19 Div under Pete Rees and the Japanese were pushed out of Mandalay on April 20.

In anticipation of the road going through the Generalissimo had ordered the move of an army of 200,000, along with a U.S Military mission from YUNAN. They crossed the Salween River in 400 rubber Storm boats on the night of 10-11 May. This force was maintained by the 13 Dakotas from the troop carrier command of the 14 U.S.A.F. This force advanced into Burma to meet Stilwell's Army.

It was on January 31, 1945 that the first convoys were flagged off from both sides. The travel time was 11 days.

It is not known how much of this famous road is still in use –possibly by the Drug Barons of the famous triangle. The Chinese have recently constructed a bridge over the river Shweli.

Bhalu in Burma

This is a story of the days when we were fighting in Burma, against the tenacious Japanese, with half trained troops, at the end of a slender Line of Communication (LoC).

The 428 Field Company of the Madras Sappers was one of the last Field Companies to be raised. Previously, we carefully formed a new Company by milking an older Company to give the new unit, a leavening of experienced VCOs and NCOs. They provided the seeds of basic *espirit de corps*, and the framework on which the unit could develop. Unfortunately, by the time the 400 series of units were raised, we had run out of the old soldiers. Each unit had to make own traditions based on the guidance from the few elders as could be scrounged from the Centre at Bangalore. The result could be disastrous—Bordie Bennet's 425 Field Company, was half its strength, with malaria and sickness, by the time it got to the front. We were lucky, after a tedious journey we reached Tengupal some 20 miles up from the airfield at Palel, on the Imphal plain. We were in support of 23 Division on the Tamu axis and we set up camp on the Khongkhan ridge, a well-drained, open and airy site.

Our Thambi's far away from home were happy except for one who missed his wife so much that one evening, he put his rifle under his left arm, and fired it. He was taken back to hospital ., we never heard of him.

Thompson, our senior British warrant officer, was a great scrounger. On one of his trips back to Palel he brought back a supply of the superior "woodbine" cigarettes, rum and tarpaulins. With trees all around, tarpaulins, tools and tradesman, we erected Swiss chalet style, comfortable, rain proof structures for our Mess—with wide roll up windows, we looked out into the wild country. On clear days we could see the Chindwin River as a silvery line shimmering in the sun, and discern Tamu in the rising mist. Most tantalising were the calls of the jungle fowl and other game in the forest below, which despite several forays we failed to bring to the table with our 303 bolt operated rifles.

At the end of a long LoC, an officer's ration was half a loaf of mouldy bread and a tin of bully beef. Matthews, the cook, had many ways to disguise the "bully", but we yearned for some "Fresh." It became an obsession we used Gelignite to stun the fish in the pools. The fish cakes were a great delicacy. We also got to know our Nagas. They lived in their huts on stilts, in the village on the other side of the road, and provided gangs for roadwork. With their rhythmic chant, *Ho–Ho*, they could shift large quantities of earth. They would offer straggly chicken in exchange for salt and cigarettes. A friendly elder taught us, how to brew *"JU"* their rice wine.

The officers, led by Captain Perdue, Crabtree, Pringle, Kochhar and Jepson soon took to playing bridge in the evenings. We were a happy lot. Rigged up a few volleyball courts where we had well contested matches. Our Thambis were great, they stood by their unit. Everyone took to mepacrin like ducks to water to avoid malaria. None wanted to be evacuated to a hospital, if sick they preferred to be treated by our own doctor Raghavan. He ran a fairly large sick bay. Just as well for him, as it came in handy when one day he had a mishap.

For our early morning business we used deep trench latrines. To keep down the smell, sweepers sprinkled some kerosene into them. Raghavan who used to smoke in the loo, one morning when he threw in a smoldering match—to his surprise and dismay—long flames lept up from the pit, and singed his bottom. Poor fellow was in agony for days—someone had poured petrol instead.

We worked under CRE 118, Colonel Richards an ex-Burma civilian. Part of a group of units tasked to`improve the hastily built road, to bring out our troops and refugees from Burma; just before the monsoon of 1942.The trek must have been an ordeal—on my first visit to Tamu, I saw metal cribs, and skeletons, all over.

A British Mechanical Equipment Company, with its dozers also camped on the ridge. Detachments of Pioneer and Transport Companies-camped at locations enroute. The task of realigning the road was interesting, but backbreaking. It started with cutting a trace through impenetrable jungle, using *dahs*. The dozers then took over and finally we built the bridges. The CRE was a prolific writer—we would receive notes such as—"I saw two sappers leaning on their shovels at MS 49.3." If we could have put these entire notes end to end, they would have paved the road all the way to Tamu.

After the monsoon of 1943, at last we got a chance to become a fighting unit. We relieved 60 Field Company, Madras Sappers and joined the famous "Black Cats".

The 17 Light Infantry Division, commanded by General Punch Cowan, the most experienced Commander on our front, had only two Brigades, 49 and 63. The Division held the high ground known as Kennedy's Peak overlooking the route to Kalemyo—through Fort White. We faced the Japanese 33 "White Tiger" Division entrenched at the old staging post, called Fort White—but there was no fort there.

Tiddim was 150 miles away from Imphal on the Manipur plain. The single lane road, deteriorated to a fair weather road after about 40 miles, till it came to the Manipur River requiring a high-class bridge. From the river, it climbed steeply by a series of hairpin bends—called the "Chocolate Staircase."

Our first task was to provide piped water supply to Tiddim. It entailed erecting seven large American Waluska diesel pumps to bring up the water from—Beltang Lui, a stream some—2,000 feet below. These pumps, had been specially provisioned from the USA, had been carefully matched, to pump up the water through a four inch victaulic pipe. Due to the poor road system, made worse by the devastating effect of monsoons, we were dependent on air supply. The greatest danger in the construction of this water supply project was due to the fact that the Tiddim end of the pipeline was at the D.Z. (dropping zone) used by the Air Force to deliver our rations and ammunition. Thus, one of the dangers of the war was from the missiles in the shape of sacks of Atta-arriving unsuspectedly from the air and knocking off the working paties.

The climb up and down was so arduous—that I recall having to use salt tablets. Bhalu Kochhar, incharge of this project, and his detachment camped at site. Success depended on the installation of seven of "Waluska" pumps of which only eight were available. These had to be lowered to the selected stages—this was done by local labour under Sapper supervision. Once when a pump was being slid down the narrow track, suddenly a snake slithered across and scared the work party from the village. The Naga girls and boys giggled and let go of the ropes, sending the pump hurtling down the *khud*! When the loss was reported to the CRE, Lieutenant Colonel Ward-MC who visited us and spent the afternoon watching the progress but didn't once mention the loss of the pump.

I went back, Bhalu arranged for him to spend the night at his camp. He told me later, that Ward talked of every thing except the accident. Next morning, he made an early start; as he was departing, he said to him. "Mistakes and accidents will always occur—they can be forgiven only once." So saying he left—with Bhalu still at the salute.

A great soldier, alas he was killed as his landing craft hit a mine entering Yangon river, and laid to rest at the Hanthawady Cemetery. When Bhalu and I went to Burma, recently we planted a special flag at his grave.

We were in support of 48 Brigade. One of its Battalions with which we had a lot of '*milap*' was the 2 /5 Royal Gurkha Rifles. Just before our joining the Division, Havildar Gaje Ghale had won a V.C. During an attack Ghale was commanding a platoon of young soldiers, neither he nor his men had ever been under fire. He was advancing from Fort White to capture a stockade (an old staging post) which had been lost to the enemy. The attack by the Company led by an officer was met with intense fire. It was successful due to the example set by the Havildar's grit in assaulting despite being wounded. The position was over run against great odds.

Our Gurkhas with their aggressive patrolling had a fairly good picture of the opposition facing us at the strong point called '*Bashas*' (bamboo structures). The Japanese had built themselves formidable bunkers dug into the hillside, with roofs of two layers of logs with six feet of earth on top. No shell or bomb in use up to that time had proved effective against these bunkers. The Sappers were called in—they devised pole charges to destroy these defences. Jepson a subaltern, who had joined as a Chepstow boy in the Royal Engineers and knew his workshop practice, took control of the production. Using plastic explosive (PEK), which had to be kneaded like dough to make it pliable, in specially designed boxes out of old *ghee* (butter) tins. These were mounted on top of sturdy bamboo poles. The Sappers handling the PEK got headache—but the contraptions were successfully made. Next came the problem of finding the people for the assault. The pole charge with the explosive box on top was a fairly heavy and awkward load. The device, besides the strength to carry it and rush it in the assault, required the skill in pulling the detonating cord and taking cover in the split second interval. Choice naturally fell on Bhalu Kochhar, the only other regular officer in the unit. Bhalu was called so, because of his size, six feet six and strongly built. He chose the best men and got ready to accompany the 2/9th Gurkhas in the attack.

The plan, in short, was to surprise the enemy. Further, as it was not possible to provide Artillery support, the enemy position was to be indicated by our infantry by firing a smoke shell and the RAF were to bomb and strafe the enemy. Before the enemy had a chance to recover, Sappers were to advance with the pole charges destroying the bunkers, to enable the Gurkhas to capture the objective. A simple plan, its execution I will describe in the words of Bhalu as told to me on his return from the attack.

"We started at 2100 hours, going down the hill to the *Bashas.* Reached there undetected by 2 a.m. and rested. At first light, we saw the Japanese in their pajamas going through their morning ablutions-they were completely unaware of our presence. We waited for the appearance of our guide plane; slowly we saw it approach our area. The leading Company Commander fired the smoke discharge signal on the feature occupied by the Japanese. It was by then broad daylight, the Japanese lookouts were alert, and they immediately hurled the smoke indicator bomb back down the hill towards our position. Within minutes all hell broke loose. The Hurricanes were over us, dive bombing and strafing our position—Bhalu was lucky to escape—a *sathi* (Gurkha) just managed to pull me into a nullah."

General Cowan and all of us were at the advance HQ at the "Saddle"–a feature overlooking Fort White. We had watched our Air Force in action. Of course at that distance, we had no idea of the mishap on the ground—we got to know of it, only after the message from the Commanding Officer. Luckily, there were no major casualties, the operation was called off. I was glad to be able to look my best friend Kumar Kochhar, Bhalu's elder brother in the eye, for not having written off his brother.

After the fiasco for a proper attack on Fort White, it was decided to establish a forward position on the route. It was to be held by the Punjabis supported by a supply point and a sub section of the sappers to look after the water supply.

I had a bright Malyalee NCO, Dorairaj Nadar, a go-getter who I had just brought down from the rank of L/Havildar, to Naik for driving a vehicle without a license. It was a difficult task, for a youngster to not only set up, and work the water point but also to look after his men, in a hostile country. I decided to try him, and am glad to say that he did not let me down, he did us proud.

The Japanese Tiger Division was a well-trained formation that had been honed after long years of operations against the Chinese. It mounted a very quick attack against us. The situation was grave, we sat at the Advanced H Q at the "Saddle", and saw our Punjabis and others all getting back through our lines. There was no sign of the Sappers. We waited the whole of the first and then a second day. Finally, on the third morning, we saw our Sappers climbing back through the mist. Believe it or not, they came up all properly dressed up in their greatcoats, (it was a chilly morning) carrying their own, as well as some enemy weapons, and equipment. It was a great relief, their delay, they said, was due to the fact that they did not receive orders to withdraw.

They, therefore, worked as per the book—sent out an evening patrol to check on the enemy activities. Unfortunately, Lance Naik Joseph was caught unawares; he was disarmed and taken to an officer for questioning. The Japanese were driving the jeeps up and down, scattering *Atta* for which they had no use. The Lance Naik as the second in command of the sub-section was armed with a Light Machine Gun (LMG), which the enemy took away, but he still had a spare barrel of the LMG, tucked under his belt. The Japanese officer was casually waving his pistol at him. Joseph took a chance using the spare barrel he knocked the pistol out of his hand and ran down the slope, keeping low—as per book—and got back to report to his Commander. Thereupon L/Hav Dorairaj started operation withdrawal. Burnt all papers, distributed food, and moved to an alternative position. They used circuitous route to get back—avoiding enemy patrols. Punch Cowan visited our lines—to congratulate the men. Dorairaj was awarded an IDSM. Later Brigadier Horsefield the Chief Engineer, visited; he selected me as his Staff Officer (Operations) at his HQ.

At XIV Army HQ

A move from the best command in the Army to a staff officer was terrible. The Headquarters, a British formation, had no drivers or orderlies. If one had to go anywhere, you just went to the vehicle park and took a vehicle. For air travel, one just rang the 3 Tactical Airforce HQ (US Air Force} next door. No forms or requisitions were required. I was the lone Indian Officer. We lived in three messes. Worked seven days a week, took turns at night duty, once a fortnight. Every afternoon, exactly at 4 p.m., Scott who dealt with personnel and I with troops and equipment, got a fixed time call to Delhi. We just had to ask what was wanted and it came ASAP (As soon as possible). It gave me an overall view of the operations. But it took me a long time to forget my Thambis.

To get to know the XV Corps in the Arakan, I went on a tour, flew to Chittagong—the blue sea and the white bungalows with gardens, were a wonderful sight. A great change from the jungles and hills. A jeep from the Corps HQ took me on the long road journey. Met Brigadier L I Jacques, the Chief Engineer, my ex Commanding Officer at the Training Battalion. I toured right forward to the Buthidong Tunnels.

Soon after my return, on February 3, at 2300 hours the Japanese launched their HA-GO operations, against us in the Arakan. Depending on surprise, the Japanese took suicidal risks. In a quick enveloping action, they surrounded Headquarters 7 Indian Division. For the second time in his life, General Messervey escaped being taken prisoner. Our troops had been trained to hold fast in administrative boxes and fight, while they were maintained by air. The two reserves Divisions from Chittagong were moved up and soon had them packing.

I lost my friend Madhav Bewoor. He was killed by a sniper's bullet while visiting his defended posts. A pity was that though I had received the bad news and was on line to Delhi; I did not have the heart to inform Sir Gurunath, his father.

The main Japanese attack, UGO, to capture Imphal followed. They attacked 17 Division at Tiddim, and 23 Division on Tamu front. Both were advised to pull back towards Imphal. The first indication of the Japanese attack was on March 9, when a two men patrol of the 1/10 Gurkhas reported a 2,000 strong Japanese force crossing the Manipur River, below Tiddim. Withdrawal started at 1700 hours on March 14. Major Siddhu, a doctor, recalls how they destroyed everything, even empty bottles, which may be of use to the Japanese. 428 Field Company, which had in the meanwhile received a British regular officer as their Company Commander, blew up the water supply installation as well.

In the withdrawal from Tiddim, 17 Division was greatly harassed by the Japanese roadblocks placed on the single way road. They played merry hell with the retreating troops. Fairly early, while coming down the Chocolate staircase, 428 Company lost Captain Perdue, the second in command, who was in the front seat of a Dodge. He was killed by a burst of MG, on his head. There was great confusion during the withdrawal. While the young officers Crabtree, Jepson, Pringle, Bhalu and the VCOs took charge and extricated the company, Jim, the new Officer Commanding disappeared. No one knew as to where he was?

Siddhu (now a retired Brigadier) recalls, how he was able to extricate his Field Ambulance, in the midst of heavy fire by the Japanese from across the river. It earned him a well-deserved Military Cross. He also told me how an enemy patrol crossed the river and immobilized a large number of vehicles of Mohite's, MT Regiment in camp they quietly removed their "Rotors." Luckily our air supply was good. A timely drop soon had them on the road again with necessary supplies.

A surprising revelation by him is that—despite the retreat by the troops, the Deputy Commissioner of the Chin Hills stayed back at his post, in March 44. He was sheltered by the people, but did not have to be in that state for long as within three months the situation changed and General Mutagachi, ordered a general withdrawal.

Imphal plateau was besieged. It was held due to the availability of Air Transport. Two Divisions were flown over from the Arakan. 2 British Division from India plus three additional Brigades were moved in to stem the Japanese onslaught.

Returning aircraft evacuated—hospitals and surplus non-combatants, to reduce pressure on supplies. Kohima, on the way back to its main base at Manipur road or Dimapur, was soon also cut-off. Some of the severest fighting of the Burma campaign took place during three months partly through the rains.

The intensity of the fighting can be gauged by the fact that defending the area our Black Cat Division won three Victoria Cross in a span of a fortnight.

The first glorious feat was by a Ganju Lama of the 7 Gurkhas. Being son of a Lama from Sikkim, he was a much larger person than normal. Those of us who have ever fired the old PIAT rifle know of its very strong recoil. Ganju was strong enough to stand it. He destroyed two Japanese tanks. The White Tigers had attacked with their usual elan in the watery dawn of June 12, as the flocks of ducks were taking off from the nearby lake. One of the Anti-Tank guns of the Artillery regiment had been destroyed by enemy fire when two companies of the 1/7 Gurkhas, launched a counter attack in the afternoon. The Japanese again came to life, put down a heavy Artillery barrage; and held us up by the fire of three tanks. On his own initiative despite being wounded he crawled to within 30 yards of each tank and destroyed two of them. The third was killed by the gunners.

A fortnight later on June 26, Naik Agansing Rai of the 2/5 Royal Gurkhas—led a section of eight men at the "Water Picket" at Imphal, to attack a bunker held by the enemy with his Tommy Gun in one hand and a Grenade in the other. Bearing a charmed life, he discharged his grenade and Tommy Gun right into the aperture—killing all occupants. His was a rare example of a V.C. being won without a scratch. It was the first time that the Japanese ran abandoning their weapons.

The same day another V.C. was won by Subedar Netra Bhadur Thapa also of the 2/5—also at the 'water point' at Imphal. He commanded and held a most difficult and important post throughout the night. In spite of rain, enemy action, enormous casualties he held it with his Khukhri in hand repulsing attacks.

Kohima was another tough nut. I quote the Royal Engineers history of the war Volume IX, to record the bravery of a Lance Naik of 2 Field Company of the Bengal Sappers. When the Japanese infiltrated into the bakery area of the perimeter. He attacked it with explosives and wrecked the building, killing many. It was during this interlude, that L/Cpl Harman won his V.C. trying to take a prisoner. The total bag was 120 killed and many more wounded.

Return To Tiddim

The final Japanese withdrawal order was dated July 10. Our veteran, 5 Indian Division, persued the Japanese 33 Division in their retreat. At the Headquarters XIVth Army, I was involved in planning the move of 5 Division to Tiddim.

In the absence of my seniors, General Bill Slim asked for my advice. We did not have enough bridging to maintain the road to Tiddim so the best we could promise was a one way road, which Bill accepted.

Finally, as the duty officer, I received the message of its capture. The date was November 19, which, checking back from my Army book 439 "Officers Record of Service", was also the day, that General Slim moved his HQ to Imphal. The 19 Indian Division under General Pete Reece crossed the Chindwin.

Fifty years later in Myanmar. The belief that a wish made at the Shew Degon Pagoda, is granted certainly worked in my case. At the end of the war, on a day's trip to Rangoon by an RAF plane I met our, comely Miss Macdonald—heart throb of XIVth Army HQ, as she a bit overfull in her uniform, sat on the footsteps of Shew Degon. I just, had time to make a wish. It was granted and after fifty years not only did I get a *Darshan,* I stayed in a room with a glorious view of the immaculate Buddhist shrine.

We, seven war veterans, were guests of the Myanmar Army. Everyone went out of his or her way to help. We met probably the world's happiest people with lighter skins, and no whiskers, jovial and friendly. It was a pilgrimage to a very clean country. We flew to Pagan (Bagan) the cultural, and tourist centre, on the Irrawaddy River. The seat of Government of 55 kings, over twelve centuries, from 1080 AD. A vast area of 16 square miles is covered by over 2,000 bell shaped monuments. Pagodas are made of solid brickwork raised on a series of terraces; these consecrate some relic or commemorate a holy place. Temples are hollow vaulted, and enshrine Lord Buddha's image. We prayed at dazzling white Ananda Temple, built to a square plan with an Indian spire. Four standing Buddhas, 31 feet (9.5 m) high above an eight feet high throne, hewn out of a single teak, face outwards. A massive building with thick walls, corridors with natural light provided by deep windows.

A great sight was the sunset from the Hetkyomuni temple on the cliffs at Nayangu—overlooking XIV Army's main crossing of the Irrawady. The longest river crossing of the Second World

War. It was mounted in great secrecy, by IV Corps, which had moved in a right hook from Kalemyo, on the Chindwin via Gangaw to Pokkoku on the Irrawaddy. That awful track which we used, has now a railway line. We motored to the Japanese Base at Meiktila. A town with two enormous lakes—with temples in the lake which looked pretty when lit up at night.

Mandalay, the capital founded by king Mindon in 1857, is laid out as a perfect square. The roads numbered serially run east to west are crossed by streets. The Fort's two kilometer long outer walls, facing the cardinal points, are surrounded by a moat. The palace at the centre of the Fort, originally of timber, with elaborate flame finials—suggesting it being an immortal's pavilion—was destroyed by fire during the war. Since then restored in concrete, it is a prime tourist attraction. Guides spin stories of the king's jewels, 59 wives—of whom four attended, at a time. Unfortunately, he didn't enjoy them for long, he ended up in British captivity at Vellore, and changed places with Bahadur Shah Zafar, the last of the Mughals.

The temple on the hill with its two staircases, having 1,729 steps, dominates Mandalay. It now has escalators and lifts. A tablet in the wall commemorates its capture by the 4/4 Gurkhas. The head monk had presented them a bell. At a reception of ex-servicemen, we met a retired captain who as a boy of 13, remembered our put up piped water supply to Tiddim. He had even heard of the incident of the snake, slithering past boys and girls of his village, which had resulted in the loss of a pump.

One of the advantages of my being a General was that the hotel gave me a very fine set of rooms. We would assemble there in the evenings—for a "Sun downer." One evening we got into a very good mood helped by Colonel Bhardawaj's duty free "Black Label" whiskey; Bhalu revealed the mystery of the disappearance of "Jim" the Officer Commanding of 428 Field Company, during the withdrawal from Tiddim.

Jim turned up for breakfast after 10 days, when the Company was in camp at Imphal. Crabtree, the senior Subaltern, told him his fortune in very strong words.

Eventually, when Bhalu was posted back to the centre at Bangalore, the news was extracted by Stella Tucker, our late Commandants wife.

Stella an extraordinary person, tall and very presentable blond, she had a jolly, friendly, difficult to describe, horsy laugh. She had a way of getting people to do things for her, no body could refuse her. The news of the deplorable scandal of the disappearing Company Commander, had trickled back to Bangalore. Angela, Jim's wife, keen to find out the truth, enlisted Stella's help. Knowing Bhalu's weakness for tennis, they arranged a game at the mess court.

After tennis followed the usual drinks. Poor Bhalu had no chance of evading the issue against the heavy bombardment of questions and had to spill the beans. The outcome was traumatic—Angela divorced Jim. This was confirmed when I recently read his obituary. He had married thrice.

Wingate & "Mad Mike"

It was three days before he was killed that Scottie and I went to have a closer look at Wingate as he walked into Headquarters XIV Army, to meet General Bill Slim. Of medium build, he looked outlandish in an ill-fitting khaki uniform. With an extra large solar hat, and an unkempt beard he looked more like Emperor Haile Sellassie. His turnout had little connection, if any, with his origin as a cadet with a "band box" cap, worn at the Royal Military Academy—The Shop. He had come to discuss the Second Chindit Operation "Thursday." Later, we heard, that it was a stormy meeting. Whereas Slim agreed to release 14 Brigade which had been earmarked for Kohima, he refused to allot extra aircraft to fly it in earlier. Wingate protested and threatened to appeal to Churchill at which our great Bill pushed across his message pad to send a signal to whosoever he liked. Slim stood his ground. The message, if any, never went.

"Operation Thursday" was a huge show involving two divisions worth of troops and eleven U.S. Airforce squadrons. Two brigades had already been landed by air behind the Japanese to "fight it out," not merely harass. A concept developed from guerilla action, into long range penetration divisions operating far ahead of the main army–of which more anon.

Charles Orde Wingate, son of an Indian Army officer, born in 1903 at Nainital, was commissioned into the Artillery in 1923. Early during his regimental service, he showed interest in Oriental studies, particularly Arabic. Took six months language duty leave and traveled in the Middle East. Served for five years in Sudan. On leave from Sudan, he met his future wife in Egypt. She was then only sixteen. Instead of going up to Oxford, she married him at eighteen.

Wingate had won a D.S.O. when he was on Wavell's staff in Palestine. He had trained the Jewish Haganah in night operations to protect the Iraq-Haifa oil pipeline from attacks by the Arab bands. Early in the war he got a chance to try out his theory of unconventional warfare—the need for armies to send spies to conduct special operations behind the enemy lines; this was a new concept of large Long Range Penetration (LRP) groups being inducted behind the enemy to give fight.

Working in Abyssinia, Wingate's Sudanese battalion and local units, together with irregulars, which had come over to the Emperor, drove 12,000 of the enemy in Gijjam to Debra Markos on the road to Addis Ababa. Half of them were taken prisoners, the rest fled north to Gonder. Finally, he escorted Hail Selassie on to his throne, after our 5 Indian Division had defeated the Italians at Alam Algi.

The Burma Front

The year 1942 had been a bad year for the Indian Army—it had been pushed out of Burma. The Japanese won by the simple device of infiltration, litter parties cutting our communications. By the monsoon of 1942, we were back in Assam—our attacks in the Arakan were not making much headway, either.

Field Marshal Sir Archibald Wavell, then the Commander-in-Chief of India, knew Wingate's genius and had been impressed by his brilliant leadership during the Ethiopian campaign. He chose Wingate to mount a long penetration expedition, deep into the jungles of Northern Burma. He wanted to destroy the myth of the invincibility of the Japanese, and prove that, with air supply the army need not depend on a road as an umbilical cord to its base. When the road behind them is cut, they would get their supplies by air.

Wingate was promoted Brigadier and entrusted with the task of selecting and training his band of volunteers. They lived miles away from civilization, trained through the monsoons in the jungles of a Maharajah's hunting lodge. They lived hard, became tough and learnt the importance to kill before being killed. They were called "*Chindits*" after *Chinthey*, the mythological beast, half lion, half griffin, statues of which stand guard outside Burmese pagodas to ward off evil spirits. A lot of myth was built round these operations. Wingate believed in a bit of mystery, he was unconventional, wore a pith "Solar Hat," went about shirtless in camp—munching raw onions and enjoining others to do the same to get hardened.

For the maintenance of the force, it was necessary to have mules. Whilst on Wavell's staff at Haifa, Wingate had become pro Jew. Possibly because of his antipathy to the Arabs, he refused to include our Muslims from Poonch, the best muleteers, in his expeditions into Burma. Neither the British, who tended to make friends with the mules, nor the Gurkhas were so good at managing them. These temperamental creatures, whose vocal cords were removed to prevent them making noise, would run amok when under fire and cause losses of equipment. Mules are strong swimmers but they caused immense delay in crossing of the Irrawady. They had to be led into the river, individually. Finally they were killed and eaten.

The troops allotted to him a Brigade strength for the first and two Division worth for the second operation a year later, were not all that special, as neither Army H.Q. Nor Bill Slim, were convinced that diversion of resources to the large extent was desirable. Long-Range Penetration groups had been used in the desert with great success but only small numbers of men were involved.

The First Chindits Brigade consisted of a British Battalion. Kings—conscripts from Liverpool area, an Indian Battalion the 3/2 Gurkha Rifles, originally raised for garrison duties on the NWFP, elements of the old Burma Rifles, who had volunteered to stay on, and a Commando Company formed round the nucleus of men trained at Major Calvert's Bush Warfare School at Maymyo. A Sapper demolition squad, a Signal's detachment and an RAF contingent provided the technical backing.

The expedition was divided into six columns, three of which, numbers 4, 5 and 6 were mainly British and three, Nos. 1, 2, and 3 were Indian. Each with a sprinkling of Burma rifles. An interesting record was published by Capt. Rao[3], who had volunteered as the medical officer of No. 3 column, commanded by Major Mike Calvert, my course mate at "the shop." A ruthless type up- to dirty tricks—even then he was known as "Mad Mike." Stocky and tough with extremely powerful shoulders, I still remember the ducking he once gave me during a water polo match. He pushed me, down under, even before I had got to the ball!

Among the colorful personalities was a monocled Major Ferguson, whom I got to know only after the war. Another Sapper officer a Scotsman, must have also looked odd, dressed in a kilt, and a bamboo hat instead of a Gurkha or Junglehat. The kilt came in handy for looking at a map under a shaded torch—its other use for lighting cigarettes under it—was discontinued after a burn injury to his delicate parts. He was particularly gifted at laying booby traps.

The RAF contingent, under Squadron Leader Thompson, which walked all the way had the most important task of arranging supplies. With their long-range wireless sets, they could get into touch with their base in India, and call for supplies. This party also helped in the selection of DZ's (Dropping Zones) and guided the planes to them.

When the preparations were complete on January 5, 1943, Wingate was called to meet Wavell, to be told that the operations were cancelled. Incensed, he argued for two hours, till his Lordship at last relented.The Chindits were allowed to strike at the

Japanese as an independent force –not part of a major offensive, by the XIVth Army.

Wavell visited Imphal to see them off—called it a great adventure. They walked in mid February, did not want to use transport on the motor road to Tamu. The expedition went past my Field Company's lines. Today, I am probably the only one alive to have personally known the main *dramatis personae*.

- The Japanese soldiers are not all that good that they were believed to be.
- Troops need not get worried about being-cut off, there, was guaranteed air supply.

It was a morale building exercise. The men were trained to advance with confidence into enemy territory. The first obstacle was the River Chindwin. The columns split. No. 1 took the southern route; it crossed one day before the others, and marched boldly to convey the intention of going straight to Kalewa. Others crossed during the next night and even the next day, no enemy was in sight. Once across the Chindwin Japanese troops were sighted. Various encounters took place. We were able to get help from the local people, as we had Burmese speaking officers. The main railway line from Myitkyina to Mandalay was crossed; portions were attacked and demolished.

Special operating procedures were evolved as mentioned below:-

Bivouac Venues—BV. These were selected before starting on a march.The intended encampment site for the next day's halt was selected off the map. The night march was usually of ten hours duration, 7 p.m. to 5 a.m.; with a halt of 20 minutes after every 100 minutes. At dawn, parties went into camp at the preselected B.V.

Dispersal: Two drills for dispersal were followed by the Chindits. The signal for the first type of dispersal was given when they did not want to get too involved with the enemy. On the sounding of this signal, all ranks were required to slowly disengage, disperse into the jungle and make for the selected rendezvous by the appointed time.

The second type of dispersal was more serious. This was sounded only when the fighting was going very much against them. Everyone was to disengage and using devious ways to get back to where they had started the march.

Wounded: For the first expedition, there was no way to evacuate the wounded that were unable to walk or ride. The system was to bury a cache of money (Rupee coins) at a spot shown to the wounded and then take him to the headman of a friendly village. The doctor would negotiate with him and place him in their care, with a gift of money and salt, to the headman. Few of our men did get back after recovery, some came back through our camp, which was on a ridge running down to the Chindwin. The Nagas wrongly called headhunters, were in fact very helpful. There is a famous tale of one Johnny Gurkha coming back, clutching on to his map, saying how valuable it was in guiding him all the way. On final questioning, he agreed to show his treasured map—it was a map of the London Underground. How the hell a map of the London Underground could get into his hands is beyond comprehension. Possibly, he acquired it when he was an orderly to some British officer.

The Gurkhas provided the fighting power. Their *Khukries* were invaluable, both in attack and jungle clearance.

The RAF signalers with their powerful sets were in constant touch with the base. Arrangements for re-supply were excellent. In fact these were very fine-tuned. Amongst the special drops were monocles for Major Ferguson. The solicitors of an officer sent him a copy of his will which he had not signed so that he could sign it and send it back with a survivor. There were other lighter moments. The snuff for an officer got exchanged for curry powder for the Gurkhas, it made a poisonous stew. The RAF never let them down; regular supplies of rum kept them in good spirits.

Camouflage and Concealment were given the utmost importance, a few points are mentioned below. Making of false track including fresh animal droppings. Walking in Chuangs (fast streams) for long advances, leaving no tracks was a common practice. Fires were allowed to be lighted only half an hour before daylight or after dusk—to prevent the telltale hanging smoke from giving away positions.

The Details: 77 Independent Brigade (Chindits) crossed the Chindwin on February 14/15, in two groups. The northern group including 3 and 5 columns crossed at Tonhe, the southern Nos. 1 and 2 crossed opposite Tamu and reached the railway line first on the March 2 and destroyed it at several places. The northern columns, travelling via Pinion on the longer route had Calvert and Ferguson, who got to the railway line on March 6. They destroyed several bridges.

When they crossed the Irrawaddy, their troubles began. The Japanese attacked them in strength. On March 27, Wingate decided to jettison the heavy loads, kill the mules and make their way back. Whereas Wingate and the main party managed to recross the Chindwin, our heaviest casualties were during the return march. We lost a thousand men. Almost one third of the strength. An important loss was Capt. Katju M.C., the Public Relations Officer of the force. He had gone forward to a village to look for boats for the crossing. He had a premonition that he would not get out alive and he did not return.

Next morning they were able to contact our troops on the far bank. They sent across boats, provided covering fire against interference by the enemy. Once across, they were home. Feasted, they were picked up by transport and brought back to Imphal where many of them needed hospitalization. They had not shaved through their two months stint in Burma—it saved ten minutes per day for precious sleep. Before embarking for Delhi, there was a rush to be photographed with their beards. A party under Major Gilkes went straight ahead and met up with the Chinese on the Burma Road. They were well received and were soon on their way to Kunming by motor transport. They were flown back to India over the "Hump" by the American Airforce.

Chindit–II Operation "Thursday" Churchill was greatly taken up with Wingate's exploits. He called him the Clive of Burma, a man of genius and audacity and sent for him. He had just got back from the first expedition. He reached London after three days flight from the front, just as the Prime Minister was to depart for Quebec. His wife was woken up in Scotland and joined her husband aboard the *Queen Mary*.

Churchill presented Wingate to the US Chiefs of Staff to convince them that the British meant business in Burma. It was at this time that Mountbatten was appointed the Supreme Commander SEAC. A very much larger operation, by the 3 Indian Division of six Brigades, lavishly supported by the US Air Force was approved. Wingate was promoted Major General, Colonels Calvert and Ferguson as Brigadiers United States Airforce; were stated to deliver this glider borne force into the heart of Burma, early in the year 1944.

British and Gurkha troops were the prominent partners in **Operation Thursday**—launched by the 3 Indian Division on March 5, 1944. Its aim was to cut the enemy's communications near Indaw, and to upset their supply system, particularly of the Japanese 18 Division which was facing Stillwell's American forces. America was committed to the restoration of the Burma Road into China. The Chinese had been fighting Japan since 1936. Strategic support of China with its massive manpower offering a secure base on the mainland was essential.

The Chindit operations looked at as a whole assisted in the capture of the North Burma, but their effect on the battle faced by the XIV Army in the defence of the Imphal plain was hardly noticeable. For this show, in the interest of morale Wingate insisted on light casualty evacuation planes. Special "Aerographs" were distributed to the families, who were asked to write. Special arrangements were made for receipt of letters and tidings of family welfare.

The order of battle of the force was as under: -

The 16 Brigade—Brigadier Ferguson—Queens and Liecesters and 2 Field Company Royal Engineers, they marched on February 5, south from Ledo—to Indaw parallel to General Stilwell's forces

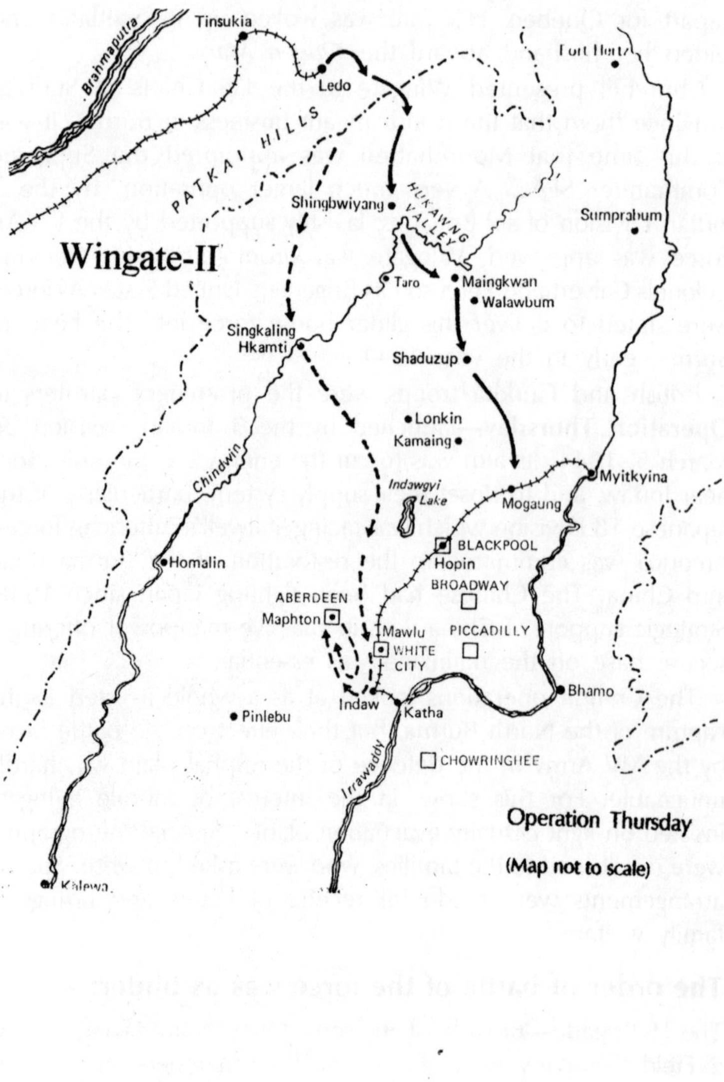
Wingate-II
Brahmaputra
Tinsukia
Ledo
Fort Hertz
PATKAI HILLS
HUKAWNG VALLEY
Shingbwiyang
Sumprabum
Taro
Maingkwan
Walawbum
Singkaling Hkamti
Shaduzup
Lonkin
Kamaing
Myitkyina
Chindwin
Indawgyi Lake
Mogaung
BLACKPOOL
Hopin
BROADWAY
Homalin
ABERDEEN
Maphton
Mawlu
PICCADILLY
WHITE CITY
Bhamo
Indaw
Katha
Pinlebu
CHOWRINGHEE
Irrawaddy
Kalewa
Operation Thursday
(Map not to scale)

- 77 Brigade—Brigadier Calvert—Kings (Liverpool), Lancashire Fusiliers
- 111 Brigade—Brigadier Lentainge—South Staffordshire and Gurkhas
- 14 Brigade—Brigadier Brodie—Black watch, Beds and Herts and Gurkhas

Of the six Brigades originally allotted, one was kept back. It was eventually diverted to the defence of Kohima. Another, the West African Brigade, was split for the defence of the airstrips.

The 77 and 111 Brigades were to be flown in, starting on March 5.The fly in was watched by General Slim, the armada took off from Lal Ghat airfield in the Assam. Just before the instructions were being given to the American pilots a photograph was received showing tree trunks across one of the landing site, called Piccadilly. Slim was present, a decision was taken—to go ahead and direct the landings to another site named Broadway.

Calvert's 77 Brigade was the first to take off in 26 planes and 54 gliders. Two gliders were being towed by one plane. This was a very heavy load—some gliders went astray and crashed. It landed at Broadway. Other landing sites were named White City, Blackpool and Chowringhee. Brigadier Lentainge's 111 Brigade landed at Chowringhee. The U.S Corps of Engineers landed early and by next day within 22 hours prepared Broadway as a landing strip. Sixty gliders, half a million pounds of stores and 1250 mules and ponies were landed. By March 11, 7,500 men with all their gear had been landed 100 miles behind the enemy lines.

The 16 Brigade, under the monoceled Brigadier Ferguson, marched through mountains and jungles, 450 miles from Ledo, solely supplied by air, caught up with Gen Stilwell's troops. The 14 Brigade was flown in on March 23 to join 16 Brigade, for the capture of Indaw. Unfortunately, the next day—March 24, Wingate was killed in an aircrash. Brigadier Lentainge, whom we later got to know as the Commandant of the Staff College, took over. For the next two months Calvert's Brigade assisted by Brodie's Gurkhas and the West Africans, held up all attacks against the White City airstrip, stopped trains and carried out demolitions on the Japanese L of C. It was called "White City," as there were a large number of parachutes hanging on the trees.

For General Joe Stilwell the highest-ranking American officer with the Chinese, early completion of the Ledo Road was vital. It is not generally known that after approval of Wingate's scheme at Quebec, he was taken round the USA. He visited American factories, where he sampled the latest foods, and also helped to train American Commandos called Merrill's Marauders after their Commander who with the help of the two Chinese Divisions Numbers 22 and 36, which had been trained at Ramgarh, fought their way through the awful Hukong Valley and captured the airfield at Myitkyina on May 17. The airfield was a great asset in the air supply mission to China. A small Japanese force however held the town. They were dislodged, with the help of Chinese volunteers.

After Wingate's death in an air crash on March 24, the command of the Chindits should rightly have passed to Brigadier Calvert. But Bill Slim gave it to Joe Lentaigne, a fellow Gurkha.

The control of the Chindits devolved on Joe Stilwell, known as Vinegar Joe. To complete his task of opening the Ledo road to Lashio, it was essential for him to capture Mogaung. For this he wanted the help of the Chindits. He had heated discussions with Mike Calvert, who pleaded that his men were tired. A medical inspection was ordered. With the end of the war in Europe and the original promise held out to the Chindits of a short 100 days' operations a large number of the British other ranks were found unfit. The Gurkhas naturally fared better. Joe distributed some American decorations.

The capture of this important railway town involved very severe fighting. With cries of *"Aye Gurkhali"* the 3/6 Gurkhas rose to the occasion, attacked and captured the railway station on June 23. Captain Allemand and Rifleman Tul Bahadur won two V.C s that day. Mogaung was cleared on June 26.

The monsoons broke at the time of withdrawal. Indwagi, a 15-mile long lake, was used for landings of two Sunderland flying boats named Gert and Daisy. Over 500 casualties were evacuated to the hospitals in India. Interestingly, one of them was Lord Wavell's son—he had asked that the more seriously wounded be given priority.

The Follow Up

The last of the Chindits left on August 27. Before the end of July, 36 British Division augmented by an Indian Brigade and Indian Sapper units in support was flown to Moguang to reinforce General Stilwell's U.S forces. 36 British Division travelled down the railway corridor from Myitkyina towards Mandalay. It received massive support from Colonel Cockran's 3 Commando U.S. Airforce. A huge formation, with 30 fighter bombers, 12 bombers—a total of 408 aircraft.The Division crossed the Irrawady unopposed but was later held up at the Shewali river. Here one of our Sappers, Rajwade won an M.C.

The Ledo road to China built by an aptly named, General Pick, was opened to traffic on January 31, 1945. The journey from Ledo to Chunking took 11 days.

The Controversy: There was a lot of criticism at our XIV Army H.Q on the resources committed to the Chindits. The diversion of resources, six Brigades equivalent to two Divisions, and eleven airforce Squadrons was unjustified as we were ourselves short of resources to fight the final thrust of three Japanese Divisions. The situation was saved by the diversion of 79 transport aircraft from the Mediterranean. This decision was taken at the highest level, at the behest of Lord Louis Mountbatten. It enabled us to reach Rangoon just as the monsoons broke.

Both Ferguson and Thompson were knighted, unfortunately friend Calvert got involved in some shady business. He won a bar to his DSO, took over command of the SAS, and finally presided over its official disbandment in 1946.

A Flashback

We old soldiers are all alike—we crave to go back to our old hunting grounds. Brigadier Lord Bernard Ferguson landed at Imphal in January '60. Permission to visit the area was denied, till he met an enlightened bureaucrat, Mr. Tayal, then the Chief Secretary of Manipur, who issued him a special permit. He became a friend of the family. Tayal's son my colleague, later introduced him to me. He found Lord and Lady Bernard (a relation of the Queen) extremely kind hosts on his visits to their estate in Ayrshire. Bernard's father was the Governor General of New Zealand—and believe it or not—he followed him in the exalted appointment. His grandfather, Sir James had also been Governor from 1869 to 1873.

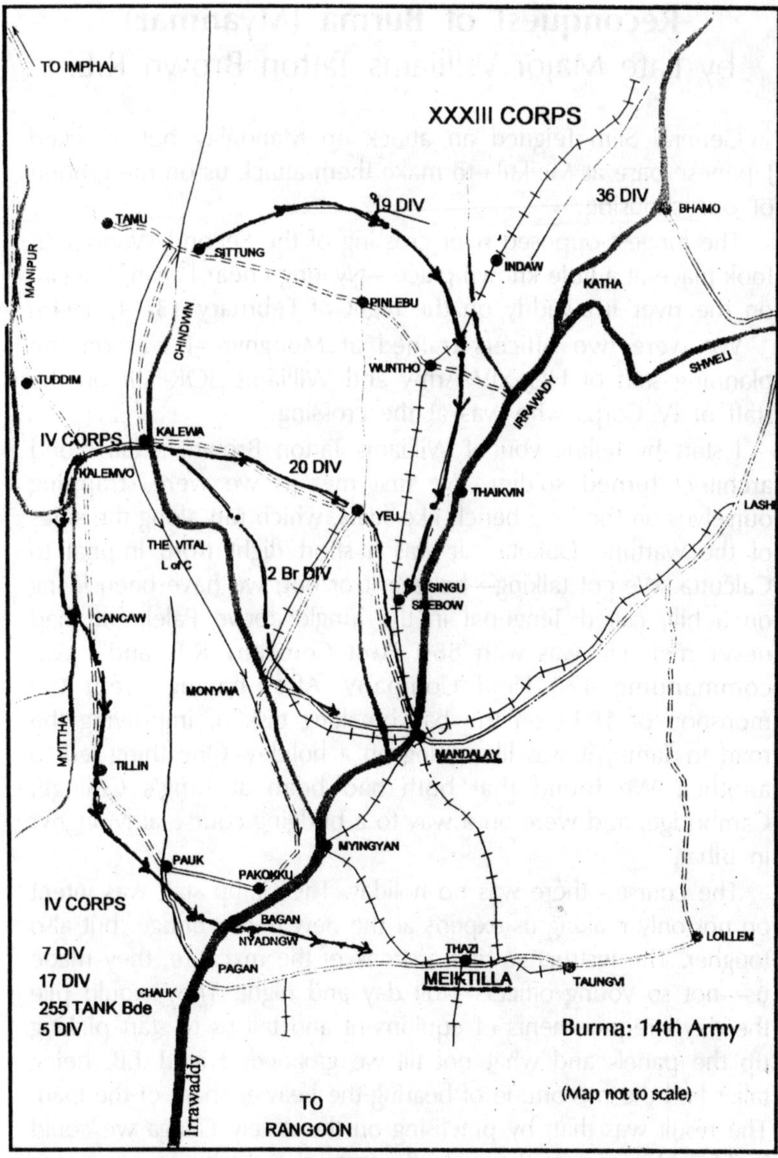
TO IMPHAL
XXXIII CORPS
36 DIV
BHAMO
19 DIV
TAMU
SITTUNG
MANIPUR
INDAW
KATHA
PINLEBU
CHINDWIN
WUNTHO
SHWELI
TUDDIM
IRRAWADY
IV CORPS
KALEWA
KALEMYO
20 DIV
THAIKVIN
YEU
LASHIO
THE VITAL
L of C
2 Br DIV
SINGU
SHEBOW
CANCAW
MONYWA
MYITTHA
MANDALAY
TILLIN
MYINGYAN
PAUK
PAKOKKU
IV CORPS
BAGAN
NYADNGW
LOILLEM
7 DIV
THAZI
17 DIV
PAGAN
MEIKTILLA
TAUNGYI
CHAUK
255 TANK Bde
5 DIV
Burma: 14th Army
Irrawaddy
(Map not to scale)
TO
RANGOON

Reconquest of Burma (Myanmar)
by Late Major Williams Tatton Brown R.E.

General Slim feigned an attack on Mandalay but attacked Japanese base at Meiktila to make them attack us on the ground of our choosing.

The longest opposed river crossing of the Second World War took place at a little known place—Nyaungu near Pagan, (Bagan) on the river Irrawaddy on the night of February 13/14, 1945.

We were two officers trained at Monghyr—I was on the planning staff of HQ XIV Army and Williams SORE III on the staff of IV Corps who was at the crossing.

I start by telling you of Williams Tatton Brown, a tall blond architect turned soldier. We first met as we were strapping ourselves on the long bench like seats, which ran along the sides of the wartime Dakotas. It was a short flight from Imphal to Calcutta. We got talking—believe it or not; we have been living on a hill, called Tengupal in the jungle above Palel, but had never met. He was with 864 Plant Company R.E. and I was commanding 428 Field Company. After having spent the monsoon of 1942 on the backbreaking task of improving the road to Tamu, it was like going on a holiday. One thing led to another. We found that both had been at King's College, Cambridge, and were on a way to a bridging course at Monghyr in Bihar.

The course—there was no holiday. The Camp staff was intent on not only making us experts at the new Bailey bridge, but also tougher. The instructors had a streak of the macabre, they made us—not so young officers—toil day and night. They would pile the day's requirements of equipment and tell us to start picking up the panels and what not till we groaned. Friend T.B. being taller had the misfortune of bearing the heavier share of the load. The result was that, by practising on the stately Ganga we could build bridges across the wide Burmese rivers even in our sleep.

I, the only Indian on the course, was lucky. My determined wife had travelled all the way from Peshawar to be with me at Monghyr. It was at the height of the "Quit India Movement." An

English lady and our Alsation in the coupe, to give them the necessary security. I met her at Kuil Junction. The Commissioner, also a King's man, allotted her a suite in the circuit house from where she watched us lugging Bailey panels. I, thus, had the bliss of going back to a comfortable bed, in her arms by night—far far away from the war.

After the course was over we went back to the jungle. My Company moved to Tiddim to fight alongside the famous 17 Black Cat Division, and I finally landed at General Slim's HQ. T.B went back to his Plant Company.

We never met till Williams, some three decades later, by another unbelievable coincidence, read the announcement of the engagement of my son in the "Times." We soon caught up and were in each other's home. On a visit to Delhi, while his wife Eilene was on a painting holiday, he wrote the story of **"RECONQUEST of BURMA".**

"My Commanding Officer, Major Gray D.S.O, was a delightful person, an architect from Dundee, but the rest of the officers were road contractors, tarmac engineers in civilian life good at their job but with very little in common with the "administration officer," which was my designation. Our troops were efficient mechanics, good drivers and hard workers who did a magnificent job in the most appalling conditions. One of my jobs was to visit them taking their mail and drink ration four bottles of beer a month. I reached after a rugged four hours drive in a Jeep at one of the camps and said "Hello Blinkinsop here's your mail and drink ration." He took the four bottles of beer and drunk them staright off, wiped his mouth and said "Cats' piss". That was all. He was not interested in his letters and didn't ask a single question about his mates. He was absolutely and completely "browned off."

On another occasion, ENSA arranged for Noel Coward and Vera Lynn (the forces sweetheart) to come to Imphal. I could not go, so I sent my Batman. When I asked him how he enjoyed it, he replied "Rather have had the same weight in beer Sir." They called Vera Lynn "Dysentery Lill" because she sang mostly to troops in hospitals where they were recovering from tropical diseases.

I found it all so depressing that I took my Urdu examination for transfer to the Indian Army. I was promoted Captain and posted to Headquarters 114 CRE in charge of the Garrison Engineers in and around Imphal. This was much more to my liking. The C R E. was an Anglo-Burman engineer, the second in command; a very nice Indian officer a most colorful regular Major, a Sikh Harkirat who smoked openly in the mess and regaled us with amusing stories. After some time I got a chance for active service. I was glad to get away. I was posted as the staff officer to the Chief Engineer IV Corps under General Messervy.

Generals and Generalship

At the Corps HQ, I had an opportunity of seeing some of the leading figures in action. These notes are offered as a "Worms-eye" view of some of the Commanders.

The most impressive was General Slim, an ex-Gurkha with a lean hard-bitten face and a prominent jaw, he introduced himself to us saying, "When I was a young officer, what I hated most was being buggered about by someone I had never seen." There was a pause and he continued, "well, you have all seen me." There was a roar of laughter and he went on to outline his aggressive strategy which will involve frequent changes of plan to keep the Japs on the move.

Lord Mountbatten also had an infectious enthusiasm that inspired all the Services. He had an informal approach jumping on a soapbox at strategic points in his itinerary and saying, "Gather round chaps." I learnt later that these soapboxes had been carefully located before hand, so that he could attract the maximum audience and that he very carefully approved their positions before visiting the troops on tours. But I do not think any worse of him for that.

General Messervy who succeeded Scones as IV Corps Commander had the reputation of placing his Head Quarters so far forward that it was frequently over-run by the enemy. On the last occasion, in the Western desert, he was taken prisoner. But, he had taken the precaution of taking off his red tabs and wearing a dirty old vest, he posed as a Batman. When asked by the Germans, why a man of his age was still a private he replied, "The Corps Commander has a grudge against me." So they left him and a few minutes later he escaped. In Arakan, his Head Quarters 7 Division, was over-run by the Japanese, again he escaped in his pyjamas! We were quite apprehensive when he was appointed and did not relish the idea of being over-run by the Japs. But he proved in the course of the campaign to have impeccable judgement in deciding the precise moment to move the "soft troops" in support and also how far to allow the spear-head of the attack to go ahead. On several occasions, he insisted on them coming back into a "lager" at night-fall rather than dash on and court disaster. The result was that we at Corps HQ always moved in just as the smoke of the battle had blown away, and on one or two occasions we found an unfinished meal left by our Jap opposite numbers still on the table

{My friend Williams didn't know that General Slim had given Messervey the command of IV Corps, on the firm understanding that he would take Brig Cobbs (a Sapper) as his Brigadier General Staff.}

General Pete Rees, ex-Rajputana Rifles, who commanded the Dagger 19 Division, had a great reputation for leading from the front, and on two occasions his driver was killed. When I went up to him for the Chindwin crossing, I bumped into him on the road and gave him a smart salute. He took out his notebook and said, "Who are you?" And made a note of my name. It made one feel extra careful to see that nothing went wrong, knowing that one's name was in his little book.

Historians have different views about the campaign. One school of thought, represented by Liddel Hart, highlighted the United State Naval victories that deprived the Japanese Armies support from their mother country. Cut off from supplies and reinforcements, "Burma fell like a ripe pear into the laps of the allied army."

Another view was that the Japanese Army had no anti Malaria strategy and that it was the *anopheles minimus* that destroyed them. The British and Indians won because they had better doctors.

These simplistic accounts are only half-truths. The reality at any rate for those taking part appeared rather different. Essentially, it was a problem of logistics. We had first to build roads, India and Burma had no land communications. Then we had to take the whole of XIVth Army over these roads with its tanks and transporters and bridging equipment for two major river crossings—the Chindwin and the Irrawaddy. We had to build airfields so that the army could be supplied by air, first by parachute drops and glider landings and then on regular Dakota shuttle runs.

The Japanese had only just failed to get into India after their surprise attacks on Imphal and Kohima. They were given no respite thereafter in their retreat to Burma. They fought for every inch of the way. A few machine-guns in bunkers could hold up a whole Battalion. The only way to neutralize them was by "earth-quakes."

This was the term used for artillery and air bombardment that could be called up at a few hours notice. These seldom scored direct hits, but the enemy was so stunned by the sheer weight of the onslaught that the infantry was able to creep up and throw grenades into the foxholes, before the Japs had time to recover.

To counter our patrols in the jungle, Japanese suicide squads would tie themselves to trees and wait for our men to pass below them. They held their fire till the leader was identified—usually a fairer individual, with a pip on his shoulder. They then shot him, they in turn were shot down by the Indian or Gurkha members of the patrol who returned to base less one officer.

It was obvious that with such resistance, Rangoon could never be reached before the monsoon of 1945. With the odds weighted so heavily in favour of defense by the Japanese.

The only way to defeat the Japanese Army was to get them to attack us so that we would destroy them on "Killing ground" of our own choosing. This was the secret of Slim's plan that I remember Messervy explaining to us around Christmas 1944.

The whole of the XXXIII Corps was poised behind 19 Division under Gen. Pete Rees; the thruster with Sherman tanks on

Diamond T. Transporters and medium artillery had crossed the Chindwin, 30 miles north of Kalewa, to capture Mandalay from the north.

Instead of following them we, IV Corps, were to make a surprise attack on Meiktilla, the huge Japanese central supply base south of Mandalay. This was so vital to them that they would have to counter-attack and recapture it is as soon as possible and in consequence expose themselves.

In order to achieve success, the greatest secrecy had to be observed and an elaborate deception plan was put into operation; first of all, there was a press conference and the BBC were informed of the imminent fall of Mandalay in the New Year "with an attack from the entire might of the XIVth Army. This was duly put, and of course picked up by the Japanese. Then wireless silence was enforced on all IV Corps signals, except for a small detachment, which was left behind, sending bogus messages in code by night and day on the Mandalay front. Additional activities simulated on these front included dummy paratroops. Dropping of "Parallax," which imitated sound of rifle fire punctuated with grenades. Acquaskills, which sent up very lights as they hit the water.

Then at night, we set off on a fortnight convoy down the Ganga valley, on the west of the Irrawaddy. The only troops that had been in this valley were the West Africans, whose existence had no doubt been reported to the Japanese by their Burmese informants. So orders were issued that the convoy should always be proceeded by troops with blackened faces. The jungle was pretty thick, reports of increased activity by these comparatively innocuous West Africans were disregarded by the Japanese; they were now fully concentrated against the impending attack on Mandalay by XXXIII Corps under General Stopford, with 19, 20 Indian and 2 British Divisions.

Main Crossing of the Irrawaddy

The approach to battle was led by 89 Infantry Brigade of 7 Indian Division, comprising KOSB (Kings Own Scottish Borders), 1/11 Sikhs, 4/6 Gurkhas and 7/2 Punjab with 400 mules led the advance of IV Corps. They started walking into the Ganga valley on December 31, 1944. Bill Slim visited them; behind them came the rest of the Division, its Sappers opening up the road, for the whole of the Corps, including an Armoured Brigade of Shermans on tank transporters. The road or the track was torturous. Bill Slim had a brush with a young Indian Sapper Noshir Grant, whose lorry with bridging stores, had broken down. He appeared out of the blue, and ordered it to be thrown down the '*khud*'. When Noshir protested, he scrawled his name on a message pad, and told him to give it to his Commanding Officer.

They somehow reached Pokkoku, on the banks of Irrawaddy on February 13. On their way to their target—the Japanese base at Meiktilla—IV, Corps crossing of the Irrawaddy at Nyaungu, the longest opposed river crossing of the war; took place after dark on February 13. The 89 Brigade at Pokkuko established elaborate traffic control, so that the unending convoys of bridging lorries could be unloaded at appropriate sites.

The main crossing spearheaded by 33 Brigade of 7 Indian Division, was also covered by an elaborate deception plan. Small boats with lanterns were launched five or ten miles above and below the actual point chosen. Complete silence was imposed and we were not allowed to test the out boat motors covered in dust after the 1500-mile journey over the mountain roads. A Naval officer with an Aldis lamp went across in a canoe and signaled the all clear from the far bank. Sappers paddled across a Company of South Lancashire under a moonless sky. Then a lone airplane flew slowly, over an hour before dawn, intended to drown the noise of the outboard motors staring up. Two more Companies followed in assault boat fitted with outboard motors, but only a few started. The Japanese Garrison had been woken up by the airplane and the waited for the boats to come within easy range before they opened fire. None of them landed, several of them were sunk. Others, which broke down in midstream, were swept and stranded on the islands in the blazing sun, some miles away laden with dead and dying. Of course, the Sappers were blamed "They let us down" was on everyone's lips.

Now that the gaff had been blown, General Messervey was not put off, he ordered the gunners to open up at 11.00 am and I remember seeing the second wave of 4/15 Punjab slowly carrying their boats to the water line—each one would have expected the same fate as the first wave and it was also different from the dash of a bayonet charge, and all those images of men going to war that I had been brought upon. These men were reluctant to die. They set off very slowly and deliberately. They all thought they would make it, but they did. By the time, we had got the outboard mortars serviced and checked and none of them broke down. The Japanese were still cowering in their slit trenches and many had been killed by the bombardment. The Punjabis who landed successfully soon finished them.

The four divisional engineers, corps troops engineers, bridging companies, engineer battalions were put under the command of a Brigadier. Three crossings were established. The longest was over 2,200 yards with island in the middle. It was a party thereafter at various crossings competing for customers. The Sappers had a field day, competing for traffic across the wide majestic river. Posters were put up "come on the Red route—Faster and safer on the Red route." There was no shortage of boats.

A shipyard had been set up on river Chindwin, which built tugs, barges and even two Naval vessels complete with guns. Elephants hauled the logs to the sawmills. Forestry companies produced cut teakwood. Tradesmen of Artisan Works companies constructed tugs, lighters, and barges, which were fitted with marine engines. This fleet, not only gave us additional 500 tons of supply lift per day from Kalewa to Myingyan—200 miles down the river Chindwin, it also provided water craft for operating in this riverine country.

Breakout From The Bridgehead

The need of the hour was to get to the Japanese Main Base, some 80 miles south at Meiktilla. General Cowman of 17 Indian Division, along with 255 Tank Brigade, crossed and raced forward. Bulldozers reached the airfield within 72 hours. Leading them was Lt. Bahadur Singh, of the Probyns' Horse. Another was Captain Sobha Chand of the Grenadiers, who ably supported the tanks. His Jemadar Mamura, a champion quarter miler, chased and caught the Brigade's first prisoner, a Japanese warrant officer, but his comfort girl unfortunately escaped. As soon as the airfield was clear, the airborne brigade of 5 Indian Division was landed. The 17 Division's air portable 99 Brigade was also flown in from Palel. General Cowan worked quickly and sealed the main roads. The Cantonment was taken on March 1. The Japanese reacted strongly, as expected. They launched vicious counter-attacks. Additional formations were brought in—the enemy was pitted against carefully prepared defenses. Night after night, the Japanese met their fate on the "killing ground " in front of the perimeter. Amphibious planes landed on the lakes.

The airfield was susceptible to enemy fire, so the RAF refused General Slim a flight to the front. He however, managed a lift with an American General in his Mitchell bomber and enroute picked up General Messervy at Pagan to watch the crucial battle. Seeing Gen. Cowan at great pressure, he called up our famous 5 Indian Division from Jorhat 700 miles away; it included Lt. Col. Lakhinder Singh's Punjabis.

During this time, Bill Slim had a chance meeting with one of our young Gunners, Sahdev Gupta, who was in the command post at his gun position; when Bill looked in at the same time on the wireless came an insistent signal "Mike Target, Mike Target, Mike Target", Gupta brushed the figure aside and carried out his fire plan. When it was finished, and he got out of his command post to get some fresh air, he saw to his horror the red hat and his Army Commander. He started to mutter apologies but Bill interrupted, "Don't bother about that my boy. If everybody worked like you, we would get to Rangoon a lot sooner."

In these operations, a little known American outfit the "Quaker conscientious objectors" displayed a different kind of valour. They flew Piper Cubs right up to the front line, landed anywhere to pick up our wounded. Flew them to medical centers in no time. That saved many lives, amputations and innumerable cases of gangrene.

After about a fortnight, the Japanese started to withdraw from Meiktilla. Fighting was equally fierce, on the other fronts. In the battle for Myingyan, 40 miles north of Meiktilla and half way to Mandalay, 7 Indian Division troops won two VCs. On March 8, 1945, when our advance to Myingyan was held up, Lt. Karamjeet Singh, Judge of the 4/15th Punjab, went forward to call up tanks on their house phone. The position was taken but again held up by the enemy in well-dug and camouflaged bunkers. He again went up to indicate the position of the Machine Gun nests, to the tanks. This time he was killed. For his abject disregard of personal safety and bravery, he was awarded a V.C. Naik Gian Singh of the 4/15th Punjab on the road won another V.C. to Myingyan, when he annihilated an enemy machine gun post.

Mandalay fell on April 20. The total casualties of the XIV Army were only 10,500.

Withdrawal and Follow-up

While all this was taking place, I had to attend Messervey's "Morning prayers." Every morning he summoned the heads of all arms and their staff to put him in the picture. The briefing always followed the same pattern. Col. Holbrock (who was at school with me at Wellington) the intelligence officer gave us the latest information about the enemy. This was an amazing performance given with the utmost confidence, a complete description of the Japanese dispositions and the changes that were taking place. It was a brilliant summary of all the sources of information coming in from aerial photos, reports of agents behind the lines, captured documents and so on. I asked him afterwards if he ever checked up on his deductions and predictions to see how often he made mistakes. He said he was far too busy preparing for the next day's conference.

Information about the enemy was followed by "Information about our own troops." The Brigadier General Staff described the actions that had taken place during the last 24 hours. Particularly encounters resulting in success with heavy casualties. These were always referred to as "the Gloucester (or Rajputana

Rifles etc.) is having quite a pary at mile 72." Then the Brigadier (Q) Quarter Master would give a description of the supply situation and what could be expected and when. The C.R.A (Gunner) C.E (Sapper) C.M.O (Medical) and RAF and any other heads of services that were present followed.

By this stage of the war nearly all of them were regular officers, highly professional and extremely competent. At the end of the conference, the General would ask one or two searching questions and then give his decision. The objectives were clearly announced with the times and dates, e.g., "We will take Crossroads by 1200 hours on April 19." In nine cases out of ten, this is exactly what happened, at any rate no one was in doubt about what was expected of him.

After our entry into Burma, we captured miles of railway lines in fairly good order. A group of British railway experts dressed in uniforms were asked to say how soon they could get a regular service going. "Six weeks to get the signals overhauled, we will have accidents if we start earlier and people will be killed." The General observed, "People are getting killed now at the front for lack of supplies. I would like Jeep trains to be running on all lines in six days." And they did.

Of course, there were crashes and accidents from time to time but everyone enjoyed playing trains. "Safety," the prime consideration of every railway engineer in peacetime, was not the first requirement.

Throughtout the compaign logistics played a crucial part. The General had to choose between the phases of operations namely:

a. Rapid advance, which required petrol.
b. Fighting major battles, which required ammunition and evacuation of casualties.
c. Consolidating—bringing in Bailey bridges engineer store, spare parts, etc., so that the forward troops could be sustained on a continuous basis over ever-lengthening lines of communication.

So much airlift required was just to keep us alive (although we were down to half rations) and very little was available for the competing demands of a, b and c. The General had to make the crucial decision and fix priorities. An officer failed to get the right load on the airlift and for the third day running, instead of badly needed barbed wire, we received crude oil to keep the dust down on the runways—his only comment was, "Oil is useful for putting on troubled waters."

At the end of the campaign, I asked Brig Boggs, the Chief Engineer why he never ticked off the less efficient formations. There was a very marked difference between the output and capabilities of the leading Field Companies and other units. He replied, "my dear T.B, every one is doing his utmost to get to Rangoon and finish the war. What would be the use of finding fault?" Boggs himself was extremely professional. He sat down and produced his appreciation for the Irrawaddy crossing in two or three hours of strenuous work. Two months later, his estimate of three days and a half was fulfilled almost to the letter. He explained later that the methodology had been worked out by Wellington and was part of the received wisdom out of the "Shop" (Royal Military Academy Woolwich).

He used to spend the day touring the countryside by air or jeep, returning at night to give orders to his staff. It was an extremely strenuous and often hazardous program. On one occasion I had to do a trip by air for him to find out if a bridge had been blown or was still intact. I was flown by a young American pilot in an unarmed light plane used normally for evacuation of casualties. We flew low and just cleared the trees—at the end of the airstrip and then skimmed along in the early morning mist. Every now and again a Pagoda would loam up to the port or starboard, just level with our wing tips. Eventually, we reached our map reference, the pilot banked steeply. I thought I would fall out of the plane, and leant heavily on the upper straps of my seat to restore the balance. I saw absolutely nothing. I was so terrified, but the pilot had fortunately seen that bridge was intact so I was able to report with confidence, "Everything Okay".

The Sappers had a very busy time receiving progress reports, translating orders into instructions and in my case making bids for air supply and allocating it to engineer stores in an ever changing priority list. We seldom got to the mess before dinnertime. On one occasion, I got there early and found a half-open bottle of rum, "What a lovely surprise" I shouted. There was a roar of laughter from the Gunners with whom we shared the officers' mess. It was the official rum ration that had been issued for over a month, but was entirely consumed by the time a single Sapper officer arrived. The gunners had a comparatively easy time as every Japanese plane had been shot out of the sky the RAF. They had also won every counter battery engagement and never lost the initiative.

The official record was compiled soon after the capture of Rangoon. These notes are only personal impressions which remain in the head after a lapse of forty years of a very junior officer who never fired a shot in anger."

Postscript

Fifty years later, I was at Bagan and remembered Tatton Brown telling me of how during the crossing he took an afternoon off, and visited Bagan the home of the thousand Pagodas on a bicycle. On a veterans' tour, we stayed at a hotel: the steps down from it led to the river ferry which was our crossing point "A".

Notes

1. Major General R.M. Rau AVSM
2. Major General M.R. Rajwade.M.C
3. Major General Rao—Army Medical Corps

❮6❯

Seagulls

What follows is a narration by Late Brigadier Y.P. Dev who was a salior before joining the Army.

I met Mohd. Khan in July 1945, when I joined the *HMIS Shillong*, as a midshipman of the RNVR—the wavy Navy. He a Khattak, from village Burikhel, had joined the Royal Indian Navy as a HO (hostilities only) rating some four years before me to escape, from the clutches of his late elder brother's wife who he had been forced to marry by his demanding father. He had been promoted as a Leading Seaman. Soon after I joined, I was made the Divisional Officer of the Starboard Watch, in addition, as the only Indian, other than the Skipper; I was nominated as the Ship's Welfare Officer.

About a week after my arrival, Mohd. Khan, who was in my Division, was brought up before me on a defaulter's parade. The charge against him was, that having been recently promoted as a Leading Seaman he found it infra dig and highly insulting, that the ship's Chief Petty Officer (the seniormost non commissioned officer) Aslam Khan, a mere PM (Punjabi Mussalman), should shout to him, "Oi Mohamad Khan, come here at the double." Leading Seaman Mohd. Khan had turned round and told him his fortune in the choicest Pushto. Well, discipline took its course and Mohamad Khan was reverted back to an ordinary seaman.

I was told that these ups and downs in his career were frequent occurrences. It hadn't bothered him much in the past, but it was different this time.

He had recently, during his last shore leave at Madras, married again, secretly, for the third time. He now needed the extra money and the status, as his new wife was the senior typist at the naval office, Madras. He did not want to face her till he was a Leading Seaman again.

As time went on, I found Khan, to be very proud, loyal and hard working. Properly motivated, he was capable of achieving wonders. HMIS Shillong, a coal burning Basset Trawler, was employed on Anti Submarine duties in the Bay of Bengal. How we were supposed to locate, chase and sink Japanese Submarines was a wonder.

After the war we came to know that not only did the Japanese Submarines, had a bigger gun, a four-inch, compared to our twelve pounders—they were faster than our nine knots, which we could achieve for short spurts only.

Our base was referred to as a Man of War, in naval parlance but our friends at Calcutta called our ship-a "*Balti*" i.e. bucket. During the monsoons, the Bay of Bengal is known to have the toughest weather conditions, possibly in the world. Cyclones are frequent occurrences particularly in July. It was during one of these rough weathers, that we got a contact, with what we thought was a Submarine. I was on the watch on the bridge and the time was about 0300 hours. I sounded "Action Stations" and woke up the captain, a Lieutenant Commander of the RNVR, and an experienced Merchant Navy Officer, who later rose to the rank of an Admiral in the Pakistan Navy.

One could feel the excitement, within the ship's company; depth charges and our only gun were manned, for our first kill. At about 0400 hours we thought we had got her location, and the depth charges were fired. Our chief steward Francis who was also on duty astern was so excited that he accidentally fell overboard. Everyone knew that Francis could not swim; without a second thought, able seaman Mohd. Khan, jumped into the seething sea after him.

The ship was on a "hunt", it could not stop, as it might have meant the Sub sinking us; instead of us getting the Sub. Apparently, we had missed the target and it was only after an hour when the all clear was sounded that we started looking for the two in the sea. By the time we returned to our original location, it was dawn-there was light enough to see. Luckily they were spotted; Mohd. Khan had Francis with him, suffering with shock but very much alive on a float. On my personal request the Captain once again promoted Khan for the act of gallantry.

After that experience Mohd. Khan was chastened, and never gave me a chance to have to send him to the 'Crows Nest' on the main mast—which was not only the highest point on the ship, it was located—uncomfortably close to the smoke belching funnel of the engine room.

In due course of time, we were both promoted, he became a petty officer, and I, a Sub Lieutenant—there was no doctor on the ship and I filled that appointment.

The war was over. In February 1946, we received orders to report to Bombay. Our first port of call was Vizag. We were berthed alongside another sister ship. The Captain was out of station; I went ashore to get supplies. The Supply Officer was a kindly old soul; he invited me to lunch with him.

I was returning after lunch, when Mohd. Khan came running to me, and whispered "Sahib do not go back to the ship, or even the Jetty. We have received a message from Bombay for a Mutiny that is scheduled for 3 pm. No damage will be caused to the ship. Symbolically, some ratings will be sent ashore to join the mutineers."

I advised him to remove all firing pins from the guns and keep them in safe custody, in the Armoury, and hoist a *"Jai Hind Flag"*. That's exactly what happened. British soldiers soon rounded up the mutineers. They were later released.

When we reached Bombay, like all the ships of the vintage of the Man of War, *HMIS Shillong* was decommissioned and sold. The officers and the crew dispersed. Mohd. Khan was demobilised in the rank of a Petty officer in early 1947.

Recently, I met his son at Madras, he was on his way to Dubai; he gave me the sad news of his passing away. Mohd. Khan's family in Pakistan and the other in Bangladesh are well settled. They say once a sailor dies, he is reborn as a seagull. I wonder where Khan is now based, and how many mates he is able to manage in his present *Janam* (Life)

The End of War

Rangoon At Last

The XIV Army got to Rangoon, just as the monsoons broke. The city dominated by the Shwedegon Pagoda—the world's biggest Buddhist monument—was attacked from three directions, a seaborne operation named—Dracula, an air dropped Battalion and a lone effort by General Slim and two Americans who wanted to be the first at the kill.

Dracula–the amphibious landing—had been mounted by XV Corps from the Arakan. Lord Louis Mountbatten had addressed the officers on, the beaches of Ramaree. A convoy of six troop ships and a number of Landing craft, protected by a naval escort, sailed on the April 27. Litty[1] and his friends of the 8/13 FFR in the LCI's were sick like dogs, being buffeted about in the pre-monsoon squalls. They were a happy battalion, commanded by Lieutenant Colonel Lewis, ex RAF; Shaukat Ali Shah was the senior Company Commander, others included Captain Fazal Rahim Khilji and Lieutenant Gurbachan Singh.

As part of XIV Army operations, a parachute battalion, of 50 Para Brigade was dropped at Elephant Point, which commanded the Rangoon River entrance. The 700 parachutists landed with their arms and cycles on the May 1, marched towards the city, when they saw the naval flotilla coming up the river. Monsoons had broken out three days too early. On the May 2 in the heavy rain and rough seas; the troops had been put ashore in their landing craft. Before disembarking they were issued fresh set of uniforms. An early casualty was the CRE Lieutenant Colonel Ward, his landing craft hit a mine.

The Scene at Rangoon

While the landings were in full swing, an aircraft flying over the Rangoon jail saw a large sign on the roof "Extract Digit—No Japs here." Hearing the report, Generals Slim and Messervy, and two Americans, took off in a Dakota to be in at the "Kill." They flew low from Pegu to Mingladoon, flying along the road at about 2,000 feet. They were within sight of Rangoon when suddenly a Japanese Anti Aircraft Battery opened up, a couple of rounds went through from the floor of the plane. Bill was fortunately not hurt, though the wireless set a few feet away was hit; an American Colonel was wounded in the leg. They just managed to get back to Pegu airfield, where quick surgery saved his life.

Wing Commander Saunders flying a Mosquito bomber, seeing the signs, landed on the airfield, and confirmed the earlier report. Thereupon the "Dracula" troops were re-embarked on their landing craft and sailed up the Rangoon River. The 8/13 FFR and the 8 Gurkhas, its forward battalions, rushed off the Jetty.

The sun was setting, there was no electricity, services having been destroyed by the Japanese. The Brigade task was to occupy the Government House and raise the flag on it. It sent two Union Jacks to the two forward Battalions. The Piffers Commanding Officer was offered a bottle of Scotch and ten days leave, as a reward. It was a very keen race, which Litty won with the help of his intelligence section. He told me, "By the time we got to the Government House it was quite dark, and it was a navigational feat to locate it and the way to the flag pole. There were no bystanders to help us. There was not even a bicycle, it was his infantry feet, which carried him, and the inbuilt guidance system, led him to the flagpole.

"The flag hoisting was a bit of an anticlimax. The Commanding Officer caught up with us and presented the bottle of Scotch. He had been preserving it since the day he had bought it from Spencers at Madras. He was carrying it in addition to the 70 pounds pack on his back. It was opened and passed around; everyone took a measure in his or her enamel mug. Festivities over—we moved into the Medical College and slept. Next morning we realised that we had spent the night amongst the half dissected bodies"

May 2, 1945, was the virtual end of the war. The same day the Germans surrendered in Italy and the Russians hoisted their Flag on the Chancery at Berlin. The war in Europe was over within the week. This came 6 days later, on May 8, 1945. Victory in Europe—VE day was celebrated with the usual "*Bara Khana*" for the troops.

For the first time India became a most important theatre of war. It was a period of plenty, weapons and materials and vehicles were available, for the asking. India became the springboard for the reconquest of South East Asia (SEAC). A huge Armada was got ready to launch a sea borne attack, by mainly Indian troops to regain Malaya and Singapore.

The ultimate weapons, two atom bombs, were dropped to make the Japanese sue for peace. An unanswered question, remained, "Was it necessary, to use the second Atom Bomb on Nagasaki?" Assuming that the Allies, keen on stopping war at the earliest—had to "A" bomb Hiroshima, there was no justification whatsoever for bombing Nagasaki, three days later.

Great damage had already been caused to Tokyo by fire on the March 9. The B29 planes of the US Strategic Command 48 of them each carrying eight tons of bombs dropped clusters of 38 incendiaries: and devastated 16 acres of Tokyo. Osaka, Kobe and Nagya were also ravaged—10,000 tons were dropped, till they ran out of incendiaries. The damage caused was even greater than by the Atom Bomb.

On April 1, the Americans landed at Okinawa, an island half way between Formosa (Taiwan) and Japan; its capture brought American bomber bases to within 400 miles. Marshall Tojo's Government fell and Suzuki took over as the premier on April 8. Peace feelers were sent out. On June 20, the Emperor himself expressed his wish for cease-fire. Prince Konoye was sent to Russia. Apparently, they did not convey it to the Allies—they had already agreed to declare war on August 8, they were intent of getting their pound of flesh to recover their rights lost in 1904 on the Sakhalin islands and Port Arthur.

Victory over Japan was celebrated on September 2, 1945. The event resulted in a rush for home leave for the British officers and men under the various schemes. To prevent congestion at Bombay—the facilities put up for mounting Operation Zipper at Devlali were converted into the Homeward Bound Depot (HWBD).

We, with a two and a half million, largest volunteer army, were left with the baby of dealing with the Japanese in South East Asia. After the surrender, we found the Japanese highly disciplined and co-operative.

Dodgie Som Dutt, my brother-in-law, told me how he took the surrender of and disarmed a Japanese Regiment of their famous "White Tiger" 33 Division—some 4,000 crack troops, with only a Company of about 120 men of his Baluchis. It went off like clockwork. Their behaviour was exemplary. Each man came forward deposited his ammunition and his grenades in separate piles, then stacked his weapon rifle in neat pile and walked straight into the barbed wire enclosure designated as the POW camp. There was not one deviation from the orders once given.

Captain Noshir Grant, the Adjutant of one of the troopships, under Colonel A S Bromfield related the scene as under:-

"We staggered up the gangways, laden like pack horses. As the Adjutant, I got a cabin to myself. In a day or two, life settled down to routine. I had the chore of the Captain's daily round. A group of six, consisting of the Captain, Officer Commanding troops, Orderly officer of the day, Medical officer, Mate and myself went round the whole ship once a day. The lavatories were the main targets. The Captain would enter first, lift the lid and say "Umh," the remainder would follow repeat the performance and walk on.

The Flotilla sailed on, we saw Penang, we still did not know definitely that there would be no opposition. On the night of September 10, we met another convoy and moved single line ahead to get through some minefields. Next day was wet and misty; we saw the coast and anchored some ten miles offshore. At about 11:30 Colonel Bromfield sent for me. He appeared to be running high temperature, he told me of a signal received from XXXIV Corps—stating that 3/4 Gurkhas would now proceed to Sumatra, but 3 Dogras would continue their journey to Port Dickson. The Signal mentioned the name of a ship Tankun that would affect this move. I did not bother to see the signal and left. Thrilled at action, issued necessary orders. Three hours later *S S Tankun* arrived, embarkation started and the troops sailed away by four p.m. Pleased with my work, I retired to my cabin.

Soon after I received a phone call to report to Officer Commanding Troops immediately. The old man was still looking seedy, sitting with his head in his hands but did not say much. I made inquiries as to whether anything had gone wrong, to which he replied, "Nothing has gone wrong my boy, except that instead of dispatching 3/4 Gurkhas, you have sent away the 3 Dogras. After all what is in a name, whether Gurkha or Dogra?" I knew then that I had goofed up, and as an Indian my career was at its end. I saluted smartly and left, I could not even bring myself to say I was sorry.

Following day, we arrived at Port Dickson. Rain fell incessantly, while we were waiting for LCT's (Landing Craft Tanks) to take us ashore. I went to bid farewell to the Colonel. He was expecting me. I said, I am extremely sorry Sir, to have let you down." We shook hands; he handed over a letter for the Officer Commading of my unit I was to join in Malaya. I was shaking in my pants, as I already knew the contents of the letter. When I joined the 337 Forestry Company I haltingly gave Major Munnings, the Officer Commanding, the letter. He read it, looked at me and said, "What sort of whisky were you serving the old man on the Ship," and handed it over to me to read. Colonel Bromfield had given me a thumping good report. He did not take it out of me for the slipup—those were the days of officers and gentlemen.

The landings in Malaya were amusing at times. Lieutenant Rau's[2] Field Company travelled on a Polish Liner, *Sobieski.* He told me that when they were transferred to LCI's (Landing Craft

Infantry), the Officer Commanding, who wanted the honour of stepping off first, came with us on the LCI. Laden with a lot of equipment and wearing life jackets, we cruised up to the coast in a holiday mood. The Sub Lieutenant Royal Navy commanding the LCI stopped the craft about 100 meters offshore. The Officer Commanding in his exuberance stepped into the sea and disappeared, at odd moments we would see his legs bobbing up. We later discovered that he had tied his life jacket too low, which gave greater buoyancy to the lower part of his body. By the time he was fished out, he had swallowed a lot of seawater. He vented his wrath on the young Naval officer. Next time, he took us into the coconut palms, and had difficulty in getting his craft away.

A Japanese welcoming party at the beach was bowing to us with monotonous regularity. Their discipline was so good that after fighting like tigers, they completely packed up, when ordered. They took us on a tour of their defenses, and gave us full co-operation. There was chaos at the Morib beaches, a few tanks gently sank till only the turrets could be seen. The sand on the beaches looked firm, but had a sub-strata of mud underneath. This had happened, despite reconnaissance and the analysis of samples before the launching of the operations. Two Dukws (amphibious Vehicles} of 25 Infantry Division were the first to enter Kuala Lumpur on September 9. Major Sahdev Gupta—(later Lieutenant General) of the 19 Division who was to land on the beaches near port Swettenham told me that when he looked at the obstacles erected at the beaches he thanked his stars that they did not have to land against opposition.

Surrender Ceremonies

The first surrender took place at Rangoon where our troops were already in position. Captain K C Khanna described the scene to me. He and Leslie Sawhney of the 18 Cavalry happened to be at Mingledoon airport and watched the arrival of two planeloads of senior Japanese officers from General Terruchi's Supreme HQ at Saigon. RAF fighters escorted these small white painted planes. The pilots opened the doors for the Generals and their staff officers. They wore smart khaki uniforms, breeches, polished leather riding boots, small peak caps, but looked incongruous in white shirts with open collars. They carried their long Sumarai swords. They were whisked off posthaste to meet Lieutenant General Stopford, who had taken over the command of the XIV Army.

The Surrender At Singapore

The preliminary surrender parade on September 3, was followed by a parade in front of the Singapore Town hall on the 12th. Lord Mountbatten, in the presence of General Slim and Brigadier K. S. Thimayya representing the Indian Army, took the surrender of all Japanese forces in the South East Asia Command. This solemn ceremony was held in the council chamber of the Municipal Corporation of Singapore. All the allied commanders were seated, even Japanese highest-ranking officers, were escorted up the steps of Town hall to the ceremony by Lieutenant Colonel Sarabjit S Kalha DSO, MC and Bar, the Commanding Officer of the 2/1 Punjab. They filled in and were seated bareheaded at a table opposite. Lord Mountbatten announced the terms of surrender—these were signed at 1100 hours. An Indian Field Battery fired the 21-gun salute. This event is depicted pictorially at the—Santoba Island about ten miles off Singapore, along with photographs of the military parade, wax models of the three dignitaries, and the sinking of *HMS Prince of Wales* and *Repulse*.

The Dutch East Indies—Java

Indonesia was our greatest test—three Indian Divisions were involved. The 5, 23 and 26, of which 26 Div-employed in Sumatra got off the lightest.

As the armada, with the Indian army aboard was still on the high seas—26 Division was directed, to send 36 Brigade to Batavia, and 71 Brigade to Sumatra to disarm the Japanese. The delicate situation of dealing with the Indonesians was explained by the liaison officers who came on board. Whereas the Dutch wished to re-establish their rule, the Indonesians who had been liberally armed by the Japanese were not willing to accept Dutch rule.

The Indónesian leader Dr Soekarno had declared a Republic three days after the Japanese surrender. There was a Dutch interim Government at Brisbane with some Dutch troops. The 9 Australian Division supported by Australian Naval and Air Forces and US had captured Borneo and Sarawak for the Dutch.

The Netherlands Chief of Civil Affairs, Van Der Plas, and Colonel Kennedy of the US were aboard British Admiral Patterson's flagship Cumberland. The Dutch had no intelligence whatsoever on the mood of the Japanese, they were expecting to be received as liberators. Mr. Plas went ashore to meet them, but beat a hasty retreat to the ship for safety.

Admiral Patterson's orders were quite clear, he was not to use the troops to suppress the Republic or recognize it. He was there to disarm the Japanese, evacuate them and the Prisoners of War. There were over 100,000 Dutch and Eurasian women and children in internment camps, who needed protection.

When the first Brigade of the 23 Division landed on September 25—guerilla bands of Indonesians had started trouble. Surrender from the Japanese at the Naval base at Surabaya was taken by a Dutch Naval Commander without adequate means to look after the arms and ammunition. The Indonesians who were then in a very truculent mood looted these weapons.

When Brigadier Mallaby's 49 Brigade got there, rioting started in earnest; first the negotiations went well but hot heads took over and Brig. Malaby was murdered. By October 4, barricades started appearing in the streets. Additional troops of 5 Division was called up to help. They landed on November 9. Without going into details, the town was brought under control after hard street fighting.

Two Indian senior Commanding Officers were Major (Later Lt Gen.) J S Dhillon of the 2 Field Company of the Bengal Sappers and Lieutenant Colonel Sarabjit Singh Kalha DSO, MC and bar commanding the 2/1 Punjab. They invariable met for lunch. One day when Dhillon went to lunch with him at his mess not finding

Kalha there, he walked into his room in the bungalow next door. Strangely, as he entered he felt as if he saw an apparition.

Soon after, Kalha's body was brought in. He had been ambushed outside the town and killed instantaneously by the burst from an automatic on his forehead. One of the finest officers was lost at the altar of colonization.

The Headquarters of the 23 Division was at Buitenzorg—now called Bandung. After the situation improved—the troops were able to enjoy the Christmas festivities. 68 Field Company Bengal Sappers under Inderjit Singh Johar started operating the printing press. It not only produced the Divisional Newspaper; but also printed currency notes "Guilders," the legal currency, which was disbursed to the officers and troops. There were parties for the children who had been in captivity for long. Dances and plays—amply compensated the troops. One of the officers told me of the system in his battalion—where except for the Commanding Officer and the medical officer—all others were allowed to use official transport for recreational purposes in the evenings. The only restriction was the strict curfew hour at ten p.m. Anyone arriving late would not only be confined to the guardroom—but would be returned to the depot posthaste.

On the setting up of the Indian Government, under Nehru, our troops were recalled. One of the ships used was *"SS Highland Brigade,"* an 18000-ton meat ship converted into a troopship. Lieutenant Sharma (Tindi) told me of the job of loading of the ship. It was completed in 20 hours, continuous work with a Japanese work party of 200. By the time it was completed, he was too tired to even attend the farewell party. His VCO Nazir Shah was not in the launch though he had taken leave to persuade his Dutch girlfriend to get smuggled on board—but failed.

At last, a Brigade group of Dutch troops started arriving. Finally on November 30, 1946, our troops were withdrawn. The men came back with very heavy packs, with bolts of silks, etc., which were awkward to carry. The custom inspectors had been instructed to look the other way.

The Dutch resisted but handed over the Government to the United States of Indonesia in 1949. The Republic was formed in 1950. The Dutch came close to defeating the infant republic—till they were forced to withdraw under American threat, to cut off aid to them in Europe.

Thailand—Siam

Thailand—the ancient kingdom of Siam, holds a special fascination for us Indians as its Buddhist culture is akin to ours. I recollect stepping over the boundary at Singora—just for the heck of it. Due to the willing co-operation of the Thai Government, SEAC decided to send only 7 Division, under General Evans to Thailand. Out of the two earmarked, 26 Division was left in reserve. An advance party of 7 Indian Division, under Colonel Kalwant Singh—HQ 14 Brigade, a composite engineer company, an officer and 40 ranks of the 1 Queens, and 100 men of the 4/5 Royal Gurkhas landed at Don Maung airport on the September 3. They were received by senior Thai officers with a guard of honour. The Commander of the organisation—Force 136 Brigade, Jacques as the political adviser went to HQ Siam Forces and arranged for conveyance of the surrender orders. These were obeyed to the letter.

Brigadier Cox, Commander 33 Brigade (Ex-Bengal Sapper), took the surrender. The formal surrender ceremony parade was held on arrival of General Evans on January 11. He accepted the swords from 22 Japanese Generals and Admirals including three Generals of Railway Corps.

Other units including 331 Field Company under Major Shenoy arrived by sea between October and December.

There were 113,000 Japanese and 20,000 Prisoners of War in Thailand. Camps were built with Japanese labour. Some prisoners put on cutting the grass on the Polo ground, using penknives, were slow. The Japanese officer in charge was sentenced to a month's imprisonment, with "bodily torture" by his Commander. Unfortunately, the last polo pony died. A major task was to repair the Don Maung airfield as refueling base for planes flying to Hong Kong, One Battalion of the Rajputana Rifles was flown in to reoccupy the Island..

An amusing incident took place on the November 11. In absence of gunners, my old unit 62 Field Company was asked to produce the signal for the commencement and the close of the two minutes silence on Armistice day. They used explosives to create the bangs, the Division HQ, hearing the explosion bangs started digging slit trenches. The trades Havildar of the Company painted an impressive Madras Sappers crest at the entrance to the Company's office. Many people would ask if it was the ALFS Headquarters?

Thus ended the 7 Division's part in the war against Japan; it left Thailand in February, to take up internal security duties in Malaya. Before leaving, it held an impressive ceremonial parade in the honour of the King of Siam—Lord Mountbatten attended; 114 Brigade under Brigadier Kulwant Singh including British troops stayed till July as part of the "British Troops Siam." This famous Division, raised at Rawalpindi and which had spearheaded XIV Army's crossing of the Irrawaddy—won many decorations including four Victoria Crosses. Finally went back to Rawalpindi, to be commanded by General Mian Haya-uddin, popularly called Ganga.

Indo-China

There were great problems in taking over Indo-China from the Japanese. The French colony had been divided along the 17th parallel, the upper portion had been taken over by the Chinese. The Japanese Supreme HQ South East Asia of Field Marshall Terruchi was at Saigon in the southern part. It was intended to fly in the whole of 20 Division, to restore the situation, but due to shortage of aircraft, only the Division HQ, a Gurkha Brigade and 422 Field Company were airlifted. The French referred to all our troops as "Les Gurkhas;" as the preponderance of our troops was Gurkhas. The name was accepted without protest.

Saigon, 34 miles up the river, was a town with wide boulevards and fine shops and churches. The French welcomed our troops but the Anamosa a warlike race was cold and distant, they also wanted their freedom. The independence movement led by Ho Chi Minh from North Vietnam was described by the British as communist trouble. Some of them had formed guerrilla bands led by Japanese officers. The powerhouse fueled by a mixture of coal dust and maize was taken over by the Sappers. They operated, and also guarded it.

The 70,000 Japanese, who had not been disarmed, refused to co-operate in reinstating the French; nor would take action till ordered by the British. The local labour had vanished; public services had ceased to operate. Sniping in the barricaded streets against the French police was rampant.

When the remainder of the 20 Division arrived by sea, it was deployed in Brigade groups in different locations and took part in fighting to clear the trouble spots. The situation around Saigon was brought under control. 422 Field Company built a raft across a river; in lieu of the bridge, which had been destroyed. Colonel Shiv Dev Verma, commanding the 16 Light Cavalry took the surrender of the 55 Imperial Division Reconnaissance Regiment, which had fought against the regiment. General Gracey, the Divisional Commander, arranged for him, six officers and two VCs to be flown over to Phnom Penh in a Japanese plane.

The Squadrons of 16 Cavalry were billeted in small habitations outside Saigon. An officer's mess was established in a bungalow. Our officers were invited to use the club, but lack of knowledge of French was a great handicap.

The Marathas of the 92 Field Company, while building a bridge to an island enjoyed the luxury of staying at the Grand Hotel with fresh linen for their beds.

On the arrival of additional French troops from Europe in November, the Japanese were interned. The French took charge, and our troops returned. First to leave on December 19 was 32 Brigade which moved to British Borneo.

422 Field Company commanded by Major R. A. Loomba came back to Macassar in the island of Celebes Southeast of Borneo. In its terrible climate they took turns for a respite at the hill station of Malino.

Indo China struggled against French colonialism, till the historic battle of Dien Bien Phu in 1954. This led to the total defeat of the French Military and in consequence they decided to quit the country. It is said that the "Vanquished left by aircrafts and tracked vehicles while the victors came riding cycles and marching on their flat feet." General de Castrie had called this war—guerre sans fronts." The United States got involved in the senseless war for the next twenty years. We, in India, hardly took any interest in this war which bled the USA till they left in 1975 after paying a heavy price in human life under serious opposition back at home.

British Occupation Forces in Japan (B.C.O.F)

I am indebted to my neighbour Colonel Sohan Singh Bedi who as the General Staff Officer at the HQ at Kure was at the centre of all the doings of the 268 Brigade.

A select force—268 Brigade, under Brigadier K S Thimayya, DSO, was sent to Japan as India's contingent, as part of the Common Wealth Division, commanded by an Australian General.

The small HQ at Tokyo was under a Colonel Maxwell of New Zealand. The staff consisted of an AQ Australian, a DAG, a GSO2 (Major) and, Bedi as the GSO3. (Capt.)

Two important Indians posted to Tokyo; were Dr L C Jain (Ex Professor Punjab University) as Economic Adviser, and Dr Pal, as the Indian judge, as a member of the War Crimes Commission. Both of them had a small staff, Jain's daughter acted as his secretary. Judge Pal had an American lady secretary, he later got an assistant from Calcutta. The presiding judge Sir William Webb was keen to deliver a unanimous decision against the War Criminals. Dr Pal, a person with very high values, didn't agree and delivered a dissenting vote. India had also decided to forego its war claims for which India earned a good name.

Under a Military Government they all depended a lot on help from Bedi, which he gave readily. Dr Jain was related to Colonel Verinder Singh, who would sometimes pay them a visit. Early on, there was disagreement between Timmy and Jain—as to which of them should represent India at Philippine's Independence Day. Delhi sent Dr Rama Rao and later Bahadur Singh of the Foreign Service to deal with political matters.

Families of officers posted to the BCOF started arriving, though Nina Thimayya did not. Amongst others who joined were Lieutenant Colonel Rajinder Singh (Dhobi), Virk and Virendra Singh's wives. Jain had made arrangements for Viru's wife's reception at Singapore. Mrs. Rajinder Singh as the Sikhni hogged the arrangements as Mrs. V. Singh!

The Indian Brigade was quartered at Matsue. Timmy as the Commander lived in great style with a dozen Japanese girls to look after him—these were NISEs, second generation Japanese born in America and spoke excellent English.

General Shrinagesh inherited them. When Kumari his wife, joined him there was a reshuffle. It was said the number one chose Naik Ghulam Mohd, the staff car driver. He was indeed a popular man; when the Indian Brigade travelled by train from Oklahoma to Kure on its return journey to India more people turned up to garland and say good-bye to Ghulam Mohd than the General.

This episode was the last time in which the Indian Army as we knew it, operated. When the BCOF was returning home some of our Muslim brethren were sent back direct to Karachi.

Fifty Years Later

The two victory days, confused even the Government of India at the time of Golden jubilee of the war. Pranab Mukerjee, the Minister for External Affairs, led a delegation of Burma veterans to Celebration of Victory in Europe, at London. Later, he ruled that the war in Burma was not India's War. A totally wrong assessment of our toil for six year plus two years after the end of the war.

Notes

1. Lieut. V.N. Malhotra later Brigadier
2. Major General B. M. Rau AVSM

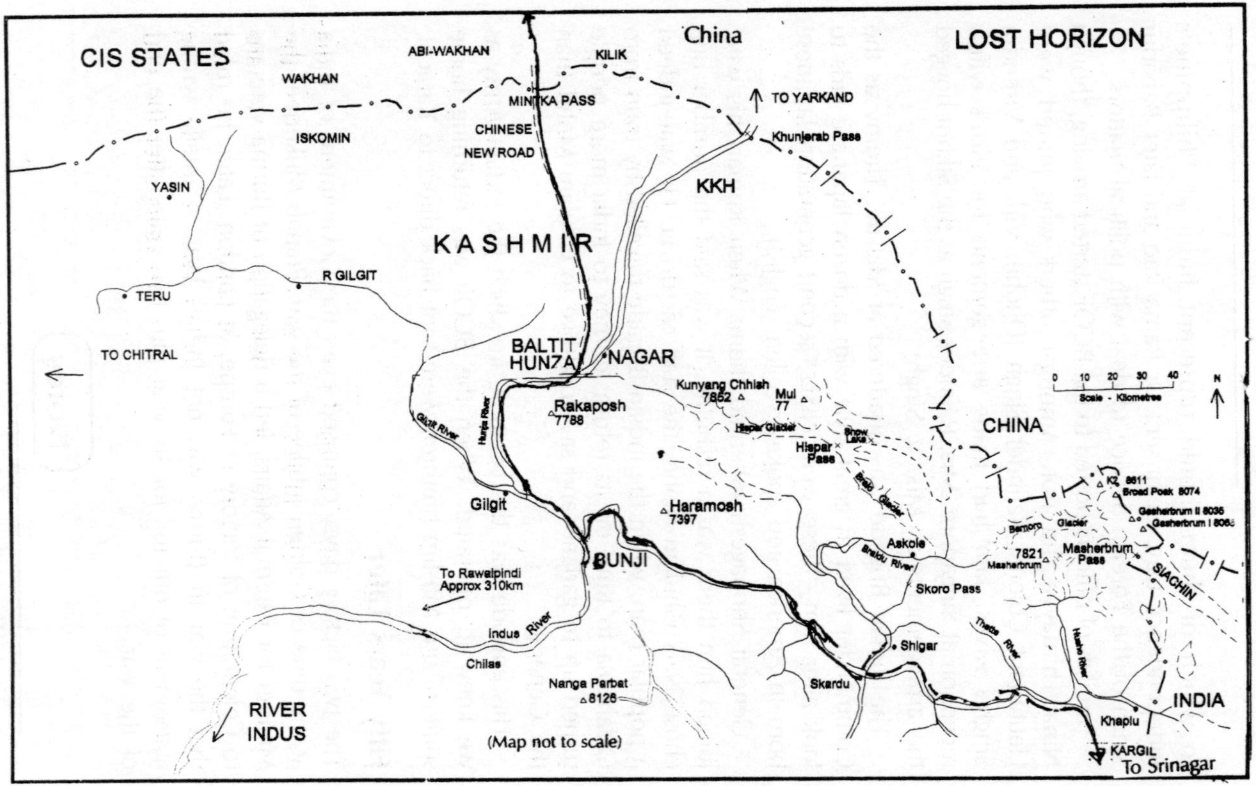
LOST HORIZON
CIS STATES
ABI-WAKHAN
WAKHAN
KILIK
China
MINTKA PASS
TO YARKAND
ISKOMIN
CHINESE
NEW ROAD
Khunjerab Pass
YASIN
KKH
KASHMIR
TERU
R GILGIT
TO CHITRAL
BALTIT
HUNZA
NAGAR
Kunyang Chhish 7852
Mul 77
Rakaposh 7788
Gilgit River
Hunja River
Hispar Glacier
Snow Lake
Hispar Pass
CHINA
Scale - Kilometres
0 10 20 30 40
N
K2 8611
Broad Peak 8074
Gasherbrum II 8035
Haramosh 7397
Gilgit
Baltoro Glacier
Masherbrum 7821
Masherbrum Pass
Askole
Braldu River
BUNJI
SIACHIN
Skoro Pass
To Rawalpindi Approx 310km
Indus River
Shigar
Thalle River
Hushe River
Chilas
Skardu
Nanga Parbat 8126
INDIA
Khaplu
RIVER INDUS
(Map not to scale)
KARGIL
To Srinagar

‹7›

The Lost Horizon

Thanks to the publicity received by Siachen, the highest battleground in the world, there is an awareness of our Northern Frontier. A little further west, are, two important forgotten lands—Gilgit and Chitral. Once the listening posts—in the vast system of National defense. During the days of the "Great Game" Czarist Russia was expanding its boundaries, the British kept a close watch on the Northern Territories particularly Gilgit and Chitral.

Realising their importance some fifty years ago it was ensured that they became part of the Western defense system. Britain took Gilgit on lease from the Maharaja for 60 years that is till 1995. A 24-40 kilometer wide wedge of Afghan territory, the Wakhan corridor, separated India, from the energy rich parts of old Russia. Just north of it, is the romantically named *Ab-i-Panja*, a stream arising in the Pamirs, to join the Oxus River.

Geographically, these lands are part of the Karakoram range, the name means 'black gravel.' A mass of rock and ice, it extends from the river Shyok, in India, for some 400 kilometers to Hunza, the northern state of Gilgit, now occupied by Pakistan.

This range is the greatest assemblage in the world—33 giant peaks, of which 19 are above 25,000 feet (7620 meters) high including the famous 'K2', the second highest peak on the earth. There were two trade routes across the Karakoram. The highest trade route of the world is across the Karakoram pass. The Han Emperors opened the Silk Route in about 120 B.C besides the Kilik Pass is a third over the Mintka Pass.

There are two separate tracks, for the summer and the winter, up to the Karakoram Pass. The summer route is open for about four months. The trade which included the import of "pashmina wool," for making "shawls," the product par excellence of the Kashmir artisans, was so important that during the Sikh Raj, a fort named Shadullah, some 125 kilometers on the north of the Pass, was built and manned, to protect the caravans. India had a Trade Commissioner at Yarkand till before Independence.

The journey is arduous but not forbidding. Recently General Vombatkere, his wife and two sons climbed the pass. They were not alone in the endeavour; they met a team of three Indian women, and a 59 year old British Army wife, who were also travelling in the area. An important post on the way to the Pass is Daulat-Beg-Oldie. Located some17 kilometers short of the Pass, it has an airfield of sorts, where our daredevil Air Force pilots could land. Interestingly, the Pass is marked by a pile of sun bleached animal bones which are easier to carry, than stones. True to the prevailing custom, they enjoyed the offering left there by the Chinese, and left their contribution for the visitors to follow.

The "Silk Route," today is a highway known as the Karakoram Highway—KKH for short. This road, built by the Pakistan Army Engineers with the help of China, between the years 1966-78, is an impressive piece of engineering as it follows the course of the river Indus, for several hundred kilometers from Gilgit. It follows the Hunza river, which flows from north to south and slashes a valley for 130 kms, through the Karakorams. It took the Pakistan Engineers, twenty years to blast it through this tough country. It cost them 400 lives through accidents and rock falls. The Karakoram Highway, following the 2,000-year-old Silk Road, runs 800 kilometers from Abottabad to the Chinese border at Khunjerab Pass, passing through Besham, Chilas, Gilgit and Nilt. Huge overloaded trucks, with the caption, "We trust Allah" traverse the highway. A couple of Bailey bridges, and a number of prestressed concrete bridges built by the Chinese, span the Indus, and its tributaries. These are well maintained by detachments, located at helipads along the road. The rigours of caravans travelling through this romantic country are no more.

A far more impressive road is the 170-kilometer long Giigit to Skardu, also built by the Pakistan Army, using Chinese engineers and labour. It too follows the Indus and crosses the river short of Skardu, by a Bailey suspension bridge. This road continues to the East, where the Indian Army Engineers too have built daunting roads over the Kardungla Pass, with the highest bridge in the world. My hope is that one-day the twain shall meet.

Some of the old soldiers will remember the gallant defense of Skardu by Major Sher Jung Bahadur Thapa of the J& K Infantry. He held it, through the winter for 183 days, till August 14, 1948. Short of every resource, he sheltered the refugees from Gigit, till it fell to Colonel Shahzaida Mata-ul-Mulk, brother of Burhan-ud-Din, the Mehtar (ruler) of Chitral. The Shahzaida sent him a message to surrender saying that in Gilgit, Hindus and Sikhs had been treated well, and he should surrender. He rejected it outright. General Chibber in his book, "Pakistan's Criminal Folly in Kashmir" relates how another message from the same Sahzaida, asking permission to kill all prisoners was intercepted by India. It was seen by General Boucher, then C in C India, who protested to his opposite number, Gen. Gracey in Pakistan. Hard to believe but true, Gracey had known Thapa, as the captain of his college hockey team, which used to play against the 1st Gurkha Centre of which he was the Adjutant. It led to our hero being alive to wear his well-earned MVC. Not only that; when a prisoner at the Attock fort, Gen. Gracey on a visit, ordered better treatment for him.

Skardu now under Pakistan occupation has an airfield to which civilian planes fly past the "Nanga Parbat," an 8126-meter (26,629 feet), massif, with an unpredictable weather. The word *Nanga* (Naked) is descriptive, as the land around it is bare and falls steeply away to the rivers. The peak in its bare glory is the most important and impressive feature in the area—looking at it is itself exhilarating.

Describing the lay of the land from the feature to the south of the Karakoram range lies Baltistan, with its capital at Skardu. To the west of it, across the river Indus are Gilgit, Hunza, Nagar, Yasin – Dardistan or the land of the Dards. The Dards are a strange tribe, very fair with European features, they have nothing in common with the Pathans or the Mongols, and have their own customs.

These lands have always had a strange fascination for most of us. Presumably, this craving for the unknown parts of the world is widespread. A Hong Kong based Chinese Travel agency offers grand travel tours—fascinating glimpses of the old silk caravan route of the romantic past from Hong Kong through Uramchi, Gilgit to Islamabad.

The way to Gilgit, which was part of the state of Jammu and Kashmir, was from Srinagar. The normal route in the old days was from Srinagar to Bandipur and then from Kargil, a 13-day trek on ponies, stopping on the way at well-appointed Dak Bungalows. On arrival at Gilgit, visitors were given a traditional welcome by a reception party beating their 'Dhols' (drums).

The history of Gilgit and Chitral, has been described very pithily by an American writer Schaller in his book "Stones of Silence" published by Vikas. The Raja of Jammu, Gulab Singh, conquered Kashmir between 1830-46, including 82,000 square miles of the mountainous Karakoram from Tibet on the east to Chitral on the west. He had some very able Generals including the Dogra General Zorawar Singh, who heaped success upon success in expanding the boundaries of the state. The trail to Leh is still known after him. When after the Sikh wars, in 1846, the British absorbed Kashmir, it had a Sikh General at Srinagar. They promptly sold it back to Raja Gulab Singh of Jammu for one million pounds, less Rs 25 Lakhs for Chamba, which was kept back by them allegedly for strategic reasons.

In 1859, the British established the Gilgit Agency. Surveys of these areas were started in 1860. The greatest peak in this area known as "K2" was discovered by Montgomery, who made the topographical survey of Kashmir and Ladakh. The first Political agent was appointed in 1877 but he left in 1881.

In 1885, Colonel Lockhart, the head of army intelligence, visited Chitral. He travelled with a sizable party from Srinagar. The Astor River had to be crossed by a rickety cantilever bridge of a 100 feet span. The Hattu Pir and Burzil Pass at 13,900 feet had to be tackled. From Gilgit, he went west to Chitral. An

Imperial figure, he went down well with the 70-year-old Mehtar of Chitral who had fathered 61 sons. Travelling further he had a bizarre meeting with the Kafirs, in the land lying between Chitral and Afghanistan. The lost tribe—they wore goatskins, worshipped spirits, and were as fair and fine-featured as the Europeans. The Kafirs had rejected Islam. They represented an ethnological puzzle; they made effigies of their dead. Their ancestors could be some of Alexander's Greek soldiers who had stayed back!

Dr. Leitner, the first Principal of Government College Lahore (1886), studied the origin of these people, their distinctive racial features and language. A remarkable person, he had earned the rank of a Colonel at a tender age of 15, as an interpreter to the British Army at Sebastopol.

At a most interesting pagan ceremony Lockhart adopted a man his own age as a son. The ceremony started with the sacrifice of a sheep, the sharing of the toasted kidneys. Then his coat and shirt were torn open, butter was applied to the left nipple and the adoptee sucked it as a vampire! The relationship was short-lived as Lockhart was ordered to return posthaste to take over as the Quarter Master General. In 1889, Algaman Durand, brother of Durand of "Durand Line" fame, then head of Political Department was appointed British Agent at Gilgit. He followed Lockhart and made an agreement with the Mehtar of Chitral. When the ruler died in 1892, there was a scramble for the throne by his 41 sons. One of them Nizam, who lived at Gilgit, was installed as the Mehtar in December 1892. Surgeon Major Robertson appointed as British Agent went to Chitral with an escort of 50 Sikhs soldiers and four officers including Younghusband, and Captain Townshend of the Central India Horse. The latter, some 20 years later as a Major General surrendered the garrison of Kut to the Turks during world war I in the state of Basra (the present day Iraq). They were besieged in the Chitral fort. Dr Robertson (later M.P), successfully held the fort, till the 32 Pioneer Battalion of Mazbi Sikhs, under Colonel Kelly, marched in from Gilgit to relieve them. The journey of the Pioneers over the snow-covered passes, manhandling guns through deep snows, was a saga of great tenacity and bravery. Unfortunately, it was not recognised as most of the attention was lavished on General Low's Chitral relief expedition, which included well-known British regiments like the 60 Rifles and the Gordon Highlanders.

In our days, Chitral ruler's brother Burhan-ud-Din was an officer in the Indian Army. According to B. L. Kak, the Mehtar of Chitral, a feudatory of Kashmir, had written to Lord Mountbatten in June 1947, expressing his desire to accede to India. Chitral—had a separate political officer during British rule.

South of Chitral were the states of Swat and Dir in the land called Kohistan. Further south through the Malakand Pass lay the North West Frontier Province.

Army Intelligence supported explorations, when Colonel Lockhart[1], the first to visit Chitral, became the C in C, he instituted the MacCregor Medal for travel and explorations, named after Sir Charles MacCregor, the Quarter Master General—as army intelligence was then under him.

The United Services Institution of India at Delhi, still awards the medal, a pity not too regularly. The large gold medal is highly prized.

In 1891 young Curzon, who seemed to have set his mind to govern India one day, visited Gilgit and Chitral, he met Captain Younghusband, who in 1904, entered Lhasa, and led our troops in a ceremonial march to the Potala Palace.

Wars had been fought between Tibet and China since the 8th Century for the control of Gilgit, because of its commanding geographical position at the mouth of the Hunza river and the Indus valley. It was a part of Kashmir State, with an area over 6,000 square miles and a population of 77,000. It had been conquered and ruled by the Dogras.

My friend General K. N. Dubey remembers being taken by his father to the Diwali Durbar of the Maharaja of Kashmir. At this annual function, not only did the Maharaja receive *Khillats* (tributes) from various parts, including gold dust, apricots, animals and what not. He even received a token tribute from Tibet. His title included a reference to his suzerainty over Tibet. Gilgit garrison was commanded by a Dogra General. Kashmir State forces, because of their frontier defense role, were allowed to hold two batteries of Mountain Artillery.

A small Hunza—Nagar Expedition in the winter of 1891, opened Hunza, the original Shangri La. The operations took place in the upper portion of the basin of the Hunza River, which flows into the Gilgit River—some two miles below the Gilgit Fort lying close beneath the mighty peak of Rakaposhi. The Gilgit River is a tributary of the Indus.

The rulers of the two states of Hunza and Nagar were called "Thumbs." It is said that it was the custom at the court for the *Wazir* to ask—in the presence of the ruler—"Who is the greatest king of the East?" To which the correct reply was, "Surely the Thumb of Hunza or the Khan of China." These two states were part of Kashmir; they paid a nominal tribute. Hunza paid annual tribute of 20 ounces of gold dust, two horses and two hounds. Nagar paid some gold dust and two baskets of apricots. Gold was panned from the sand in the river Gilgit.

A small guard of Infantry had been established at Gilgit, at the British Agency. This garrison was strengthened, and a force was sent up to Chalt in the shadow of Rakaposhi. Thereupon an insolent message was received from the Thumb of Hunza. He threatened to cut off the heads of the intruders saying "the fortress of Chalt is more precious to us, than the strings of our wives' pyjamas!" Only 1,200 men were available, including the forces and a detachment of Sappers under Captain Aylmer.

The attack on the fortress of Nilt on December 2, with over 14 feet high, eight feet thick walls perched on a cliff, with built-in towers, was a difficult operation. Captain Aylmer showed the way in. After a preliminary artillery bombardment, two guns were advanced within 250 yards of the fort. The garrison gave a warm reception; there was loud jeering and beating of the drums. The assault was uphill so they were breathless, when the party suddenly met a deep cut in front of the fort. Here with another officer, a dozen Gurkhas and sappers Hazara Singh and Abdullah, after a short break, rushed at the gate into the outer courtyard. Abdullah, with an axe and the Gurkhas with their

Khukris attacked the gate. A Gurkha was killed, and another was wounded. Hazara Singh providentially had carried a load of guncotton; he assisted Aylmer to lay a charge at the foot of the gate. At first, the fuse would not light, it had to be cut and relit. The explosion was most satisfactory, not only was the main gate blown, a number of enemy were also killed. More reinforcements arrived and the day was won. Captain Aylmer won a VC, our Sappers and some Gurkhas were awarded IOM's.

After the fall of Nilt, the enemy offered stiff resistance to further advance. Next to be captured was fortress of Thol—resting with its flanks on glaciers. It was taken on December 20, by scaling a 1,500 feet cliff, and rushing the enemy's sangars with bayonets. The ruler of Nagar was killed, and the towers of his fort were blown. The Thumb of Hunza fled to China. In March 1892, his pliable half-brother, Nazim Khan was installed as the Mir.

When in 1935, Russia took over virtual control of Chinese Sinkiang, Gilgit was taken under the Government of India because of its strategical importance. It entered into a sixty-year agreement with the State of Kashmir, and administered it directly under the Political Department at Delhi. The position was that the state retained its sovereignty, exercised through the Resident to the state. Only the administration, instead of being under the Wazir-ai-Wazarat, appointment held by the father of Colonel Shayam Rattan, was under the Deputy Commissioner. Kashmir State forces continued to provide the garrison troops as on Imperial Service with special dispensation to own Mountain Batteries.

The Assistant Political Officer in 1944-46 was my friend Jagan Nath Dhamija, a Cambridge tennis blue. He had joined the Royal Indian Navy during the Second World War. Subsequently, he was taken into the Political Department. He as the Deputy Commissioner of Gilgit and States of Hunza, Nagar, Punmial, Yasin, Ishkoman and Chalt, was called Mulki Sahib. He learnt the local language "Shina" which is akin to Kashmiri and dispensed justice as per their own tribal laws in which there was no capital punishment. A person, if he saw his wife and her paramour in a compromising position could kill both and escape free. The maximum punishment he could award was 14 years hard labour. The Resident at Srinagar confirmed the punishments.

Once in the case of a Sikh killed by another, he was asked by the Political Agent to try the case under the Indian Penal Code. He followed the procedure, found the man guilty of murder. He was sentenced to death. There were no lawyers, in the area; the mother sent a wire appealing against the sentence to the Viceroy. The file was sent for and the sentence was quashed as Dhamija, as the executive head was also in charge of the police. Those days Jagan Dhamija was a bachelor, his mother had accompanied him. They lived in a large bungalow on the river with a fantastic view of "Buddha" carved on a hill. They had all the conveniences and hordes of servants. One of the problems of living surrounded by high peaks all round, was that one never saw much of the sky—over a time a malaise called "Giligilittis" developed—a desire to see the sky.

There were exotic places to visit. He went on tours and endeared himself distributing quinine to the malaria-ridden people. He was particularly smitten by the Rakaposhi and enjoyed just sitting and watching it, the shining mass of this mountain pervades the Hunza Valley. It appears to be barring, all access to the outside world. It was the Westerner's dream of the Shangri-La.

There was a legend about a local mountaineer called Funio, who having climbed the Rakaposhi, had seen fairies. Jagan interested in things metaphysical went there on tour. After a stiff climb, he came upon a lake where it was said, one could see fairies. He camped there for two days. The rarified and calm atmosphere was certainly conducive to strange vibrations.

He left Gilgit on being appointed an Under Secretary at Delhi. When the Partition of India plan was announced on June 3, 1947, the Political Department retroceded Gilgit to the state of Kashmir. The news was received with jubilations. Brigadier Ghansara Singh of the State forces was sent as the Governor without any backing of troops to take over. The Gilgit Scouts opted for Pakistan. It was under British Officers. The Battalion of the J&K State Force, half Muslims and half Sikhs, under Lt. Col. Majid Khan, was at Bunji, 35 miles away. At midnight of October 31, the Governor's residence was surrounded, he was arrested. Major Brown of the 1,000 strong Gilgit Scouts hoisted the Pakistan flag. Non-Muslims were largely liquidated, some managed to escape to join the State Forces garrison at Skardu. Two weeks later, a Pakistani Pathan officer, Sardar Mohammed Alam, flew into Gilgit, and took over the political agency, on behalf of Pakistan Government.

My friend Colonel Lawrie during a recent visit to Hunza, the original Shangri La, saw the gun presented to the Mir of Hunza by Lord Kitchener in 1904. Kitchener had trekked almost a month from Srinagar for the visit. At Gilgit, he witnessed a six a side ferocious polo as of yore. It is said that the game originated here. It is played on a long but narrow field, by six a side teams. Originally the ball was the walnut of which trees abound; the sticks were fashioned from the branches. The best ponies came from Badakshan, which was in Russia; they were short, very comfortable and followed the ball. Children start playing on their donkeys. The main result is that lots of collarbones are broken.

River Hunza, which divides Hunza in the east and Nagar in the west, is a very fast-flowing and treacherous stream. A Political Agent and his wife, who tried to travel down it from Hunza to Gilgit by a rubber folding boat, lost their lives in the attempt. Nagar receives more sunshine but the people of Hunza are tougher, and have longer life spans. The people vénerate the Agha Khan. They follow the Ahmediya religion—which is not looked at with favour by the Pakistanis. There is an abundance of fruit trees, particularly apricots and apples. Grapes are protected by tying cloth around the bunches. They make an excellent but rather potent wine. It is said that the Mir of Hunza died of heavy drinking. Their women are very fair with blonde hair, one can get married but the girl is not allowed to leave the territory.

A war like tribe, their main source of income was raiding the caravans on the Silk Route. Their activities extended to pillaging the caravans to Yarkand from India, through the Karakoram Pass, over—three hundred kilometers away.

The capital Baltit lies in the shadow of the high, white-washed Tibetan style fort, with narrow windows. It was the seat of rulers of Hunza. In the seventies, the Mir, was an elderly man with a hint of Mongolian blood at the corners of his eyes. The state is now a district. This has happened in all the erstwhile states according to General Gul Hassan's Memoirs. The erstwhile ruler, like the ruler of Dir, and his son were whisked away in a helicopter.

The women are fair with blonde hair, well built and sturdy, they work hard in the fields, and go out daily collecting wood. They do not observe purdah. Their beauty has smitten many a visitor. Jagan told me that one of our Indian Army doctors wanted to marry one, but it did not work out, as the girl was not allowed to go out of the territory.

This brings me to what can be done today. India and Pakistan, both part of a single geographical unit, have to live and work together.

The unsolved problem is how to help Afghanistan, and open communications to the CIS Republics, to import hydrocarbons from the Caspian region reported to be floating on oil. The construction of the KKH, and the 170-km road from Gilgit to Skardu at the junction of Shoyk River with the Indus towards our road systems has already been mentioned.

The Chinese have built a new road from Gilgit to Tajikistan. What remains now is an improvement to the road from Skardu to Khaplu, near our border. We can then have a road right through from Chandigargh to CIS states. The distance across the Mintka pass via Gilgit, Skardu, Zogila pass, Banihall tunnel, Jammu, Pathankot, to Chandigargh would be about 650 kilometers. The pipeline would be shorter; the length through Chinese territory will have to be added.

We would be able to supply much needed wheat to our friends in Afghanistan. Both Pakistan and India would benefit from imports of oil and natural gas.

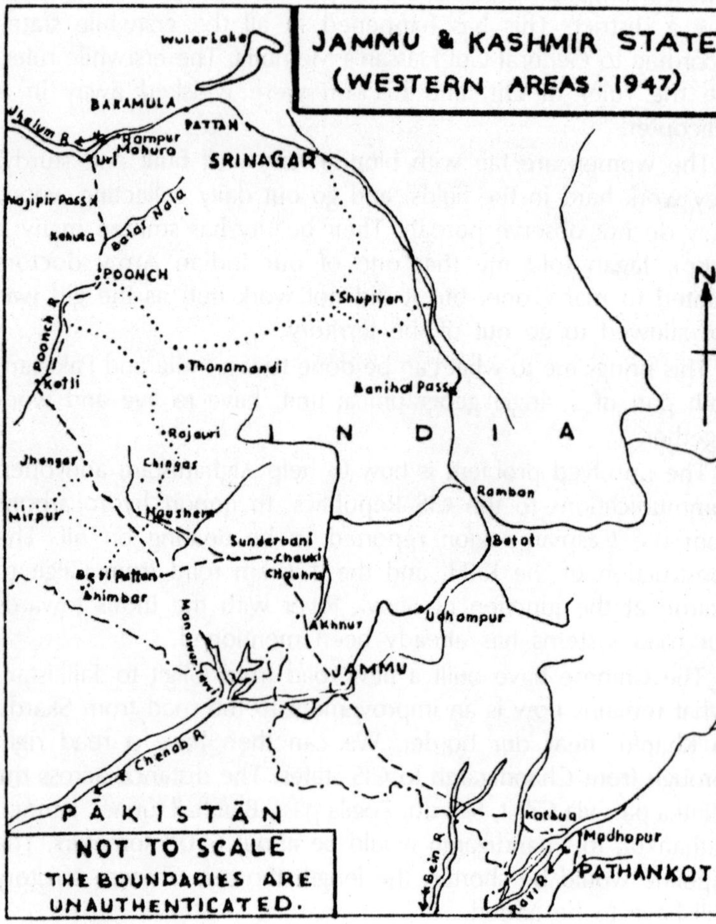
JAMMU & KASHMIR STATE
(WESTERN AREAS : 1947)
Wular Lake
BARAMULA
PATTAN
Jhelum R
Rampur
Mahura
Uri
SRINAGAR
Hajipir Pass
Batar Nala
Kahuta
POONCH
Shupiyan
N
Poonch R
Kotli
Thanamandi
Banihal Pass
Rajauri
I N D I A
Jhangar
Chingas
Ramban
Mirpur
Naushera
Sunderbani
Chauki
Batot
Bhimbar
Chaura
Akhnur
Udhampur
JAMMU
Chenab R
PĀKISTĀN
Kathua
Madhopur
PATHANKOT
Bein R
Ravi R
NOT TO SCALE
THE BOUNDARIES ARE UNAUTHENTICATED.

Kashmir Story

Here is a map of western Kashmir, scene of the operations during 1947-48.

My first vignette of some eighty years ago, is of my tiny foot reaching out to a root of a tree on the bank of the Jehlum. Trying to retrieve a stick, which had fallen into the river, next moment, I went tumbling after it, felt a sensation of fish swimming around me and then oblivion. Soon after, I found myself in the arms of the *mallah* (boatman), he brought me to the houseboat deck, shaken but sound.

People from times immemorial have been enamoured of Kashmir. The Mughals had the longest innings of some 150 years, starting with Akbar. Lovers of gardens, they left us Nasim, Nishat and Chashma Shahi. Of the three ways into Kashmir the most picturesque is the Mughal route. My parents and I once trekked on the Royal route. After a hot and dusty car ride to Bhimber, we walked up the bridle path to the Valley. I still remember the landmark of walking ten miles at the age of ten. We stayed at "dak bungalows." There were *serais* all the way up to Shupayan, the entrance to the Valley. The prettiest town on the way was Rajouri; it had some of the old *havelis* with ornate stonework. We carried our tents but pitched them only once, just before crossing the Banihal Pass. This halt, Gau Gachhi, was dismal, it rained that night—It was cold and wet. The Gujjars camping there were extremely helpful. A very fine and likable people, tending their flock, gave us milk and butter for free.

Crossing over the Pass was an exhilarating experience—a vast expanse of rolling grassland, neatly mowed by the grazing sheep and cattle. Valleys covered with banks of bluebells, daisies—a riot of colour. At the summit there were some stone cairns and flags fluttering in the wind. The gradients were generally easy, but I remember my '*ghorawala*' pointing out a slippery slope, where apparently, elephants used to slip and get killed. The Mughals also used this route to take snow back to their palaces in the hot plains. The entrance to the lush valley at Shupayan was an unforgettable experience. Apple trees on both sides of the road, one variety called "Bikh" was so plentiful that even our horses refused to eat them. Walnuts, almonds, cherry and plum trees, no wonder, it was called "Janat-ay-Benazeer".

We spent many summer vacations at Gulmarg, the home of Golf. The people were very friendly, not that orthodox—they mingled freely with us. Their women did not observe "*Purdah.*" We would go down the hill to "Baba Rishi", the tomb of a Muslim Saint—his grave in the middle with Jos sticks burning—it had railings around it with pieces of thread which wafted in the gentle breeze. People went round it, even the English tied a thread, hoping that their wish would come true.

We visited Matan, the source of Jehlum and Amarnath caves for pilgrimage, where a most wonderful tribe of Pandas (priests), Kashmiri pandits, kept records of visitors' families, through generations. They also served excellent lunches, with "*Kadam Ka Sag,*" and paid return visits to their *Jajmans* (clients) in the winters. Pahalgam on the Lidder River provided excellent trout fishing. Its bazaar sold chat and gourmet meals, which were very popular with tourists.

Partition And Rape of Kashmir

For Accession of Kashmir, to India or Pakistan—the Maharaja of the largest Indian state (2,23,870 Sq km) could not make up his mind. He entered into a Standstill Agreement, with Pakistan. Lord Mountbatten visited Srinagar for three days, also sent Lord Ismay on a fishing holiday but to no avail. The Maharaja delayed his decision, till Pakistan organised invasion by tribal Lashkars to annex the state by force and had over run Baramula.

Kartar Dubey[2] told me the real reason for delay by Maharaja Hari Singh. He had as a Guru—the famous Swamy Sant Dev, an impressive personality who lived at Kud. The Guru had fallen out of favour, but later in 1946, when the Maharaja was returning from a visit to England, his plane, forcelanded at Karachi. Hari Singh escaped unhurt, and dreamt that during the landing he was in the arms of the Guru. The Guru was immediately reinstated in favour, and accommodated at the City Palace. In return, he fed the Maharaja with dreams of his great and glorious future with his kingdom extending to Lahore and beyond. Kartar also told me of the time, when as a boy, he would accompany his father, the Governor of Jammu and Kashmir State till 1929 to the Durbar at Diwali. He saw the Maharaja receiving 'Khilat'—an annual tribute—from the Dalai Lama at Lhasa, of two goats and shawls. The Maharaja thus already had visions of trappings of greatness.

No wonder, he did not listen to Lord Mountbatten who spent three days with him in September trying to persuade him to make up his mind either way. The Maharaja feigned "Colic" to put off the final meeting. At the same time, he also deftly put off visits by Jinnah. Mountbatten was worried as there were almost a hundred British families in Kashmir, which needed protection. Colonel Dykes and his wife were killed by the raiders at Baramula. Eventually, he sent Lord Ismay on a fishing holiday but without any result.

Operation Gulmarg By Akbar Khan

I met Akbar Khan of the 6/13 Frontier Force, again at the 5th Staff College course at Quetta. He was one of the *Chalak* (clever) high flyers, married to Naseem, daughter of the influential Lady Shah Nawaz. He had connections at high places, a fertile brain, and a nascent Pathan cunning. At the time of the partition Akbar was in the Weapons & Equipment Directorate. The politicians wanting to force the Maharaja to accede to Pakistan, approached the Army for release of rifles to create disturbances in Kashmir. Akbar grasped the opportunity with both hands. He put on the mantle of the great General Tariq, Master of organisation, he sponsored the diversion of the Pathan tribesmen who influenced by the Frontier Gandhi had demanded Pukhtoonistan.

He organised 20,000 tribals in *lashkars* of 1000 tribesmen from each tribe, commanded by a Major, supported by a Captain and 10 JCOs from the regular army. After mobilisation in the first week of September, they were issued arms and ammunition and moved to Abottabad on October 18. The invasion started two days later. The orders for "Operation Gulmarg" were issued by demi official letters marked "Personal Top Secret" signed personally by the C-in-C of the Pakistan Army to the political agents. One such letter fell into the hands of Major Kalkat, who was then the Brigade Major at Bannu. On repatriation, he showed it to the Defence Minister.

The immediate concern of Pakistan was that their road and rail lines between Lahore and Rawalpindi lay within striking distance of Kashmir. To provide depth to their lines of communications, an irregular tribal force of 500 in green and khaki uniforms plus 300 Sattis from Kahura and Murree were infiltrated on September 2.

Their orders were to capture and annex as much of the State territory as possible. They attacked from Bhimber to Rajouri and on to Poonch. The massacre of the Hindus at Rajuori estimated at 40,000 was the worst. Jammu and Kashmir State Forces of three brigades under Major General Scott were hard put to deal with these tribal incursions. Major part of the J and K State forces were thus committed in penny packets. Pakistan also cut off supplies of petrol, foodstuffs, ammunition and even currency to the state dependent on it even for its post and telegraph as per the Stand Still Agreement. General Scott left about the same time as the Prime Minister Kak, who resigned on August 16. General Janak Singh took over as an interim Prime Minister. The Maharaja asked for the services of Colonel Kashmir Singh Katoch MC, to head the State forces. Katoch told me that Sardar Patel had him over to dinner, and agreed to loan his services for three years, with a provision that if Kashmir opted for Pakistan, he would revert. He reported to the C-in-C, Sir Rob Lockhart, also a Piffer, who declined to give any advice. He went to Srinagar as ordered but considered himself too young to head the Kashmir Troops, he therefore asked the Maharaja to nominate someone senior like Brigadier Rajinder Singh to head the State Forces. Meanwhile he stayed on and helped in the defence of Poonch and later in saving Srinagar.

The Attack—October 22

The State Forces were unable to cope with the situation, some Muslim troops, deserted en bloc. To defend the valley, Brigadier Rajinder Singh of Kashmir State Forces was able to muster only about 200 men. He fought a heroic rearguard action but could not demolish the main bridge at Domel, but destruction of Uri bridge delayed the tribals for a few days. He was awarded the first ever Maha Vir Chakra posthumously.

Eventually, on receipt of a telegram from the Maharaja for help, V. P. Menon accompanied by Colonel Sam Manekshaw, flew to Srinagar on October 26. They found the city plunged in gloom, the Maharaja moving from room to room with jewellery strewn all over. Menon advised him to leave Srinagar, and having accepted the formal Instrument of Accession signed by the Maharaja, they returned to Delhi.

Next morning, Sam was accompanied by the Chief to attend the Cabinet meeting at Teen Murti House. After some decision by Pandit Nehru, who wanted to ensure the Maharaja's recognition of his houseguest Sheikh Abdullah, Sardar Patel took charge—He asked Nehru—you want Kashmir? And told Sam Manekshaw, you have your orders. It was only after that the 1st Sikh, commanded by Lieutenant Colonel Ranjit Rai, were flown to Srinagar.

Kashmir Katoch met Ranjit Rai on his arrival at Srinagar. They discussed and agreed that the only way was to secure the airfield and to prevent the capture of Srinagar. The need of the hour was offensive action against the tribesmen at Baramula. The plan worked with the invaluable help of Bakshi Ghulam Mohd. who hastily rustled up a motley array of transport to lift the two companies and section of mortars of the Sikhs to the front.

Akbar Khan had promised the tribesmen the riches and "*Houries*" (damsels) of Kashmir but forgotten the golden rule "pillage and lack of discipline will invariably lead to a rout". That is exactly what happened. The tribesmen got busy in arson, loot and rape of nuns at the convent at Baramula, thus giving Indian troops time to deploy.

A gallant attack by the Sikhs on the tribesmen then loading lorries to send back the loot saved Srinagar. Colonel Rai and his men continued to fight a rearguard action. About halfway back he got killed at about 11 pm. He was awarded a posthumous MVC for exceptional bravery.

On the other route into Srinagar via Gulmarg, the tribals offered a large sum of money to a local to show them the way. However, he ran and informed the police. Major Som Nath Sharma's company of the Kumaon regiment out numbered 7 to 1, stopped and routed this lashkar at Badgam, by inflicting on them very heavy casualties. He too paid for it with his life, and won the first ever Param Vir Chakra, our highest decoration for bravery (PVC) but saved the airfield for landing of troops by IAF.

On November 1, when Lieutenant Colonel Harbaksh Singh, Commanding Officer of the Sikhs reached the airfield, he saw a battle raging at its edge. He noticed some Sikh troops unloading a 3.7-inch mountain gun, under the command of 2nd Lieutenant Jabar Singh of Patiala state. He ordered him, who incidentally was his nephew to disregard all previous orders and fire open sights at the raiders. This was very effective. Major Chachchi of Bengal Sappers flew in with two havildars to instruct the Sappers of Kashmir State Forces. Finding none were available, he came back and returned with a section to Uri for demolition. The first Sapper unit to arrive was R. N. Kumar's 30 Para Field Company. The Brigade Commander Bogie Sen[3] asked the Sappers to build a bridge across the wide and turbulent River Jehlum at Baramula. There were no stores and equipments. He threatened to send a signal to Army HQ. saying, "Two Sapper officers Chachchi and Kumar—say they cannot build a bridge." The situation was salvaged as they made a diversion to reach the existing bridge while he captured Uri on November 13.

Sardar Baldev Singh and Sardar Patel accompanied by Maniben visited Srinagar on November 4. They met Brigadier Sen and Harbaksh Singh—the situation was indeed grim—Somnath Sharma had been killed the day before. The Sardar promised and arranged Bharat Airways to fly and sent the reinforcements which were invaluable.

General Rudra told me, soon after 15 Aug., 1947 the Commander-in-Chief presented a blueprint of the future Armed Forces of India to Prime Minister Nehru. He said, "Rubbish, total rubbish. We do not need a Defence plan. Our policy is non-violence. We foresee no military threats. Scrap the Army as the police is good enough to meet our security needs."

We knew Panditji was liable to lose his calm at times. I had a taste of it myself. On a Sunday morning in the autumn of 1951 a meeting had been called in the PM's office to—discuss the floods in the valley. We the Engineers waited while Panditji and Shiekh Sahib, the PM of Kashmir, turned their backs and looked at the wall maps to decide how to control the floods. At last they turned round, Panditji told me to send troops to dig a channel on a vaguely pointed out line. I just could not hold back, and said—"Sir, don't you think this needs a technical solution?" He lost his shirt and burst out, "Women and Children will dig it etc" and walked out of the meeting. It is to the credit of Vishnu Sahay, the Secretary Kashmir affairs, who within half an hour telephoned me to find the solution. The solution was that the outfall of the river Jehlum from the Dal Lake had silted up and had to be cleared.

Poonch

Pakistan had attacked Poonch very early to gain maximum territory. There was a motorable road between Uri and Poonch over the Haji Pir Pass. After the visit by Sardar Patel, when Sen's Brigade was set to advance on its axis for attack on Domel, which would have closed the *Gate into Kashmir* forever, was directed to go to the help of Raja of Poonch. Obeying the political instructions, only one of his battalions the 1 Kumaon under Lieutenant Colonel Pritam Singh M.C crossed over. The rest of the Brigade returned to Uri sector but a golden opportunity was lost.

Lieutenant Colonel Pritam Singh was promoted Brigadier. His defence of Poonch in the face of enemy firing from the hills all round him was an epic performance. He was helped in his endeavours by the legendary Air Commodore Baba Mehar Singh[4] of the Air Force. The only way to supply was by air. An airstrip was made, and Baba Mehar personally flew in the first Dakota.

His greatest feat was to land 25 Pounder Field Guns along with ammunition and crews at night on a kutcha airstrip. The Air Force wanted 15 Dakotas to help in the airlift but the Government failed to get them.

Relief Of Poonch

Without going into the details, highlighted is the gallantry of Brigadier Usman, MVC of 50 Para Brigade, who along with Brigadier Yadu Nath Singh's Brigade captured Jhangar on March 17 and advanced on to Rajouri—the pretty town which had been turned into a graveyard by the raiders. Pakistan deployed a Division of its regular troops on the hills covering all approaches to Poonch. The tenacity of Lieutenant Rane in clearing mines, which had been layed by Pakistan without any marking along the main road, was the most outstanding performance.

It is of interest that when the year-old siege of Poonch was lifted on November 21 and Skardu was lost two days later, Pakistan agreed to accept the United Nations plan for a ceasefire. The ceasefire timing so upset Thimayya that he went on leave in sheer frustation. The Army was not consulted in fixing the timing for cease-fire.

Jammu

The summer Capital, with a railway connection to Punjab, was the most important entry point into the state. Somehow this had missed the attention it deserved. It was only a chance visit of General Dubey that the use of this entry point was accepted. Friend Dubey[5], posted as CRE 104 Engineers, was on an inspection visit to Jammu in his Dodge truck with four Sappers. He found the city in a disturbed state with sporadic firing going on in various localities. Next morning, he ordered a MG burst at an offending house. It became all quiet immediately. He went to enquire the reason for the firing. The neighbours said they were scared. The Garrison at Jammu, short of resources, could not provide any help but Kartar with his Sappers removed the rails at Naga Railway Bridge over the Balol River, to prevent the raiders using trains for attacking the state.

The Line of Communication

Our greatest handicap in fighting was, the poor state of road between Pathankot and Jammu, with three unbridged rivers-Ravi, Ujh and Basantar. From there to Srinagar, the Ramban Bridge-over the Chenab was so eaten by rust that a rod just could be put through its transoms. It took convoys five days to drive to the valley from Pathankot. We had neglected this road in spite of the fact that we held huge quantities of Bailey bridges, which were being dispatched by trainloads to Pakistan.

There was only a single runway at Srinagar, which was likely to break up in the winter due to the traffic. To keep the dust down Sappers sprayed water and managed to get some road rollers released from Sheikh Abdullah. Fortunately, we had Air Marshal Elmhirst as the Air Chief. Under his guidance an alternative airfield was developed, and as the Chief Engineer to the Air Force, I was deputed to study the RAF airfields in Europe to improve surfacing of this airfield and keep it operational through the winter.

Ladakh

Located at the crown of India is the abode of our Ice cap. This land of Buddhist culture, amidst a plethora of Muslim lands is in a way anarchism. How it has continued to retain its original faith; is almost a fairy tale. Briefly stated, in the year 1600, Ladakh was attacked and it lost the battle to the Muslims from Baltistan, whose ruler Ali Mir instead of wielding the sword of Islam, for reasons unknown, restored the King Janyang of Ladakh to his throne, on the condition that he marry his daughter, and that their son should succeed him on the throne. This was accepted and, the son Sennge Namgyal—(Sennge means Lion) became one of the most famous kings of Ladakh. He built the nine storey high palace which dominates Leh. The Buddhist faith continued to prosper. The Sikhs and Dogra Kings—through the exploits of General Zorawar Singh—tried to conquer Tibet; they captured Garo, but were prevailed upon by the British to make peace.

Through the winter of 1947-48 Pakistani forces based in Gilgit, invaded our small garrison at Skardu under Major Thapa. Short of food and ammunition, even batteries, he could only establish wireless contact, in the evening with Major Malhotra, Staff Officer to General Thimmaya . The Division itself short of troops, sent a relief column of a J and K Battalion. It failed to get to Skardu as it was ambushed half way; and returned having lost the supplies. When questioned the Commander reported, "*Addhy te Katte Gae, Addhe Bhagge Gae!*" (Half were killed and half ran away). There were troops galore in India. Skardu was lost for a while. Major (later Brigadier) Sher Jang Bahadur Thapa, the defender of Skardu, won an M.V.C for his dogged defence of the fortress. A desperate situation arose, the enemy advanced, and set up a Machine-Gun post firing into Leh—they were out to capture the whole of Ladakh. A little known Ladaki—Chhewang Rinchin, a 9th class student, along with his friends destroyed enemy Machine-Gun post on September 19 displaying great courage. His courageous and heroic deeds were duly appreciated. He was awarded, a Maha Vir Chakra.

We owe a great debt to General Thimayya, Air Commodore Mehar Singh and the lowly Pioneers of Border Roads—for Ladakh remaining with India. On May 24, 1948, a date and event which needs to be recorded in letters of gold, Mehar Baba and Thimmaya flew machine guns and crews to Leh. They landed a dakota on a hastily prepared highest airstrip then. The engine was not switched off and it flew back safely, thanks to the flying skills of our legendary Mehar Baba. The only DSO and MVC of the IAF.

Hark

We must be careful in heaping all our praise on these two luminaries. Little known to most, greater kudos should go to Major Hari Chand of the 2nd Dogras, who hailed from Leh and played a major role. His battalion commanded by my friend G

G Bewoor reached Srinagar in November 1947. On arrival, he was greatly dismayed to hear the plans of the people of Leh to migrate to Tibet. He went there in spite of the hazards of crossing the snow bound Zojila and other passes with 15 volunteers and Sonam Norbu the Engineer—who built the airstrip. His determined party hoisted the National Flag at Leh on March 13, and thereafter organised its defence by training the local youth.

Another great adventurer was Major Hari Chand of the 2/8 Gurkhas, who marched his company from Manali to Leh—an arduous journey of 42 days over three passes, the highest being the 16,000 feet Baralacha La.

Tanks on the Zojila Pass

The operations on this vital road were also beset with a problem of lack of troops. So much so that the 1st Patialas had to vacate some of their forward piquets during the winter of 1947, only to be regained at a great cost later against the machine gun nests at Zojila Pass. The use of tanks was Timmy's brainwave. He not only conceived the unusual operation, he provided the leadership to accomplish the objectives. Thangaraju[6] road, named after the officer commanding 13 Field Company of the Madras Sappers over the 12,000-feet high pass, was hewn out of rock and snow in sub zero temperatures. The dismantling of the tank turrets, bringing them forward all the way from Baltal, rebuilding the Stuarts required a high degree of mechanical skills–the tasks were achieved. The storming of Zojila by the tanks of 7 Light Cavalry under Lieutenant Colonel Rajinder Singh 'Sparrow' took place on November 1. It was a short and swift battle. The enemy could not for their life imagine that there were tanks against them at this height—they scooted.

In Retrospect—we forgot the Duke of Wellington's dictum—*"A great country can have, no such thing as a little war."*

We fought this war with our hands tied behind our backs. We had troops galore in India, yet 1 Patiala had to give up its posts due to lack of men. This glorious battalion won 7 out of total 11 MVCs and 11 Vir Chakras. Very commendable but isn't it too demanding from too few. The Air Force asked for a dozen

Dakotas —even these could not be commandeered. Waged by an unprepared nation, without any strategical thinking and ability to employ its defence forces the J and K operation came to a close with a loss of almost 2/3rd of the area of the J and K state.

The public was not even aware that there was a war. It was a comic situation. I remember going to Bangalore on a tour, and seeing Pakistani flags flying in the Commercial street.

This war was almost like an exercise with troops with live ammunition, fought under the directions of British C-in-Cs on both sides. Colonel H.S. Kapur, the ADC to General Gracy, told me how he made many trips to Lahore for the meetings of the C-in-Cs.

The "Defender of Poonch" Brigadier Pritam Singh MC, was accused of some misdemeanor involving the Raja of Poonch was Court-Marshalled—according to some in a witch hunt. Our veteran journalist, Khuswant Singh wrote in the *Hindustan Times* that in consideration of his wife's anguish, the case should be reopened. I agree with him.

Notes

1. General Sir Robert Lockhart C-in-C 1898
2. Lieut. General K N Dubey
3. Lt General L.P Sen, D.S.O
4. Air Com MEHAR SINGH D.S.O
5. Lt General K.N.Dubey PVSM
6. Maj Thangaraju—who was in Massawa-End of Italians story.

❮8❯

Midshipmen in Korea
with Rear Admiral Y.N. Inamdar

Convalescing at the Army Hospital, Delhi Cantonment, I was glad of the company of our mariner—Admiral Inamdar who was recovering from a hernia operation. One evening when I went to his room he had a visitor, an old colleague, whom he had not met for a long time. When the caller left, he said, "Well here is a strange story in which I was involved 40 years ago with the person you saw just now."

"We were midshipmen under training with the Royal Navy. The Admiralty had an excellent system of giving young officers an opportunity of, wherever possible, being trained in Fleets of their choice. The trip from Plymouth to London was, however, a hazardous journey for us Indian cadets. The distance is 210 miles, whereas the railway warrants, under travel regulation (India) were only valid for 200 miles—presumably we were required to jump off the train and walk the ten miles.

Taking advantage of the privilege, I sought an interview with the Admiral (Personnel) and asked for a posting to the Far East Fleet, the furthest part of the world. He agreed.

The two of us were appointed to, *HMS Triumph*, based at Malta which was sent to reinforce the garrison at Hong Kong. We were to join a frigate *HMS Mount Bay* in the Sea of Japan. The Triumph, sailed from Glasgow. It had nearly a hundred 'Fire Fly' and 'Sea Fire' planes on board. Being an old type of carrier without an angled deck, during the exercises enroute, there were a number of accidents. Several planes landed in the sea but the accompanying destroyers would proceed at lightning speed and invariably rescue the pilots.

The attrition rate was so high that by the time the aircraft carrier reached Hong Kong, all the planes had to be replaced.

Refueling at Aden and Singapore, we reached Hong Kong where we were welcomed with great publicity by the press. The beautiful Chinese author Han Suyin's book, *"A Many Splendoured Thing"* had just been released. In which many a Hong Kong personalities were recognizable.

Passing a bookstore, we saw a photograph of Chou En-Lai with her and her husband, Colonel Ruthnaswamy. They had met, when she was writing her book *"The Mountain is Young"*, and he was building the first road into the forbidden Kingdom of Nepal.

From there we joined *HMS Mount Bay* under the command of J.H. Unwin DSO, DSC and Bar for our training cruise. Finally, when we were in the Sea of Japan, almost at the end of our first training period of eight and a half months, the Captain decided to visit Omonato the famous Japanese Northern Sea Base for a week or so, to survey the port. On the last day the Mayor of the Town invited the ship's company for a farewell banquet. All the officers including the doctor attended the party—the Captain and we two midshipmen stayed on board. The party got back about eleven p.m., but later complained of food poisoning. They had been served best of the Japanese food, raw fish, shellfish, etc., which possibly did not agree with them.

Everyone including the doctor was violently sick.

The Captain was in a great quandary, to give up his sailing program would have meant loss of face. He talked to us, his two midshipmen, saying that as we had already gained the rudimentary knowledge, which only needed to be put into practice, he was going to sail as per his original schedule with us sharing the watches. Our instructions were to disturb him only in the case of a dire emergency. With the ship's company still sick, the doctor took two days to recover and then gradually managed to cure the rest of the officers. The ship meanwhile sailed merrily.

The Captain was very pleased with our performance. We were constantly on the Bridge for 6 or 7 days doing four hours on and four hours off. He asked me, as to how he could help us. We were then not likely to get home for almost six years, so I asked for ten days home leave. Believe it or not as soon as the ship berthed at Hong Kong Captain Unwin went to the C-in-C Far East's office and obtained not only permission to send us on leave, but also to travel all the way by air. A very special privilege those days.

When Naval HQ India was informed of the arrangements, it objected to two young midshipmen being authorised to travel by air. To which the Captain's reply was curt and final, "If Naval HQ India objects, the officers are to be withdrawn from training forthwith!" A copy of signal was endorsed to the Admiralty.

"No more was heard and we travelled to Singapore by British Overseas Airways Corporation (BOAC). Our tickets had a rubber stamp endorsement; "In case the connecting Quantas flight is delayed, passengers will be accommodated at the Raffles Hotel at Singapore." For those not knowledgeable about the Far East, Raffles Hotel is world renowned for luxury and offers the best in the way of the Old World charm. It is also atrociously expensive!

The flights from Australia were generally late up to three days. We saw prospects of luxurious living, when another snag cropped up. On going through the customs we were asked to declare the amount of currency held by us. Just before leaving Hong Kong we had been given 50 pounds each as advance of pay and allowances, we had no time to obtain travellers cheques, and were carrying cash. The rules of the Federation of Malaya were strict; the maximum currency allowed to be carried by a passenger was five pounds only.

The customs authorities threatened to send us to jail for contravening the currency regulations. However, before taking the drastic step they asked if we had a contact in Singapore. I, thereupon, gave the name of Sir Hugh Pugsley, the Flag officer,

Malaya Area. He was the father of one of our shipmates. The old man came down personally and bailed us out. He also offered to put us up at *HMS Terror*, the Royal Navy shore establishment. We declined the offer, saying that as the arrival of the Quantas plane was uncertain, the *HMS Terror*, was too far away. We preferred to wallow in the luxury of the Raffles.

When the Quantas plane to Colombo arrived three days later, we got our next big shock—we were presented huge bills for the stay in the luxury suite. We had only stayed at the Raffles, the premier hotel of Singapore, on the basis of being guests of the airline. There was no possibility whatsoever of our being able to pay the bill. This was obvious to all. The issue was referred to the Naval Headquarters, and we were allowed to continue our journey to Colombo, Madras and anon to our homes!

The BOAC like other airlines have their system of inviting the suggestions from their passengers. On the return flight on entering the aircraft, I took one of these forms and wrote off a complaint about our embarrassment at Singapore and forgot all about the incident, till one day, when I was back at Dartmouth, I received a letter, signed personally by the BOAC Chairman, apologizing for the inconvenience caused to us, due to the nondeletion of the rubber stamp on our air tickets.

Two years later (in 1952), when I was due to obtain the watchkeeping certificate, I again opted for service in the Far East. I got passage in *HMS Vengeance*, another aircraft carrier, on its way to Singapore. I was to join *HMS Belfast*, which was then operating in Korea. The war was on, and the exact location of the ship was changing daily.

At Singapore, after a spell at the Naval Air Station "*Sembawang*," I was put on board a Dakota, which with its bucket seats was not all that comfortable. We flew for four days, from Borneo through the Philippines, Hong Kong, Ewvakuni, and on to a base near the 38th Parallel. Here I boarded a Royal New Zealand ship, *HMNZS* Rotoiti, which was then under command of a Royal Naval officer.

Finally, I joined my ship, from a Chinese junk. It was my first initiation to war, and to the ancient land originally called Chosan, "the land of morning freshness."

HMS Belfast performed some daring feats. It fired over 8000 six-inch shells, in anger with such great effect that it became the subject of envy by all other navies. The fire effectiveness was the result of the Captain's special "**War Dance**" drill. This in short was entering a port suddenly at great speed, navigating in a circular path and keep on firing full blast at all targets. Someone asked the gunnery officer about the success of his shoots. He replied tongue in-cheek, "*Borrowed a spirit-level from a builder on shore.*"

In addition, it landed its Royal Marines on board, at various times for ground operations. At one such landing, they suffered very heavy casualties as the Americans had—without their knowledge—mined the area. To prevent such recurrences a liaison officer each from the US and South Korean army were posted to the *Belfast*. The South Korean liaison officer shared my cabin. The operations were so strenuous that the ship had a week off, after six weeks; alternately, at Miyajima in Japan and Hong Kong.

During an off period—I was again lucky to be granted ten days home leave—with authority to travel by air. This time though at Singapore I stayed at *HMS Terror*, while waiting to catch the BOAC flight to India. Unfortunately, I again got into trouble. This time I lost my wallet. I had put it carefully in an inflatable attaché' case, it was a clear case of theft. I immediately reported to the Dockyard police. The Eurasian Police Sergeant stopped and questioned the sailors going ashore. He narrowed down his inquiry to two stewards—one a Singhalese and the other Chinese. After questioning, he let the Chinese go out and put the other through further grilling. On a search of his locker, the money was found. I heaved a sigh of relief. Thanking him profusely—I asked him as a matter of interest, "How was it that you let go the Chinese?" He said because "A Chinese under stress would laugh and that man didn't."

While I was away the ship was attacked. A steward, who used to look after me, was killed by shrapnel fire. When the Chinese People's Volunteers entered the fray, they pushed back the UN forces and the front finally stabilised at the 38th parallel. The British Army's Middlesex Regiment organised a Tattoo on the mainland, the '*Belfast*' had some special relationship, with the regiment. Some officers were invited to a memorable evening show. I met the famous Lieutenant Colonel Rangaraj, Commander of the Indian Field Ambulance which had earned great praise for its work. Lieutenant Colonel Rangaraj and Major Bannerjee who had been awarded a *Vir Chakra* were quite obviously very popular.

The *Belfast* on decommissioning was chosen to be converted into a museum. Today it lies on the embankment of the Thames at London.

Korea is not too well known to us Indians. An ancient kingdom, it was ruled by the Yi dynasty from 1392 to 1910. Japan annexed it after the Russo-Japanese war. Its ancient name *Chosan or Choshnu*—stands for "The land of the morning Freshness." People who have lived there rave about the mornings. At the end of the Second World War, the peninsula, which had been colonized by Japan, was promised independence by the allies. It was taken over by the Chinese in the North and the Americans in the South and divided along the 38th parallel.

When the Americans troops had withdrew, the North Koreans at 4 a.m. on June 25, 1950, crossed the border and almost overran the South Korea, as the defenders were unable to stop them. The South Korean capital, Seoul, 25 miles south of the 38th parallel fell. The UN promptly declared North Korea as the aggressor. Truman, the President of the US; authorised US naval and Air Forces to aid South Korea in a "Police Action on behalf of the UN." The immediate American troops available in Japan had become soft, living on 10-cent drinks, and the company of Geisha girls. General MacArthur used US Marines, to mount a most successful landing at Inchon on coast about 20 miles from Seoul on September 15. Assault on Seoul started on September 22, and the next day the North Korean army disintegrated under American onslaught.

President Truman met MacArthur at Wake Island and awarded him Fifth Distinguished Service Medal. ROK troops, including South Korea Woman Corps, advanced across the 38th parallel to Pyongyang; the North Korean capital was captured. South Korea was supported by the troops of the 16 nations including Turks and Thais, which caused gigantic logistic problems. Britain formed a Commonwealth Division; India declined to send combat units, but provided a Field Ambulance. *HMS Belfast* was a part of the contribution by the UK.

UN forces pushed them right back almost to the Chinese frontier. When the UN forces were about 40 miles from the frontier, the Chinese sent their forces across the Yalu River in support of North Koreans. They used bugles, gongs and whistles for communications, which along with their mass attacks unnerved the defenders. UN forces push forward was finally halted on November 25. The Chinese using Russian T-34 tanks made rapid advances. In the intense cold, a general withdrawal was ordered on November 29. The US forces retreated from Hamhung, the North Korean industrial town by sea. In front of Seoul ROK troops broke; the Chinese overran Seoul and threatened even Pusan. When on March 21, 1951, General MacArthur, proposed using an Atom Bomb to stem the tide, he was relieved of his command by President Truman in April 1951, General Ridgeway took over.

The Americans with some 40,000 UN troops from Britain, Canada, etc., and the South Koreans, fought back tenaciously and eventually regained Seoul, and pushed the North Koreans beyond the 38th parallel to Pyongyang, the capital of North Korea.

On December 2, 1952, General Eisenhower, the President-elect, personally visited Korea, met his son, and his regiment. He gave a fresh impetus to stopping this senseless war. He insisted on Syghman Rhee towing the line. To persuade the Chinese that War was not profitable—power stations on the Yalu and Pyongyang the capital were bombed on June 23, 1952. After the death of Stalin in March, Chou En-Lai agreed to a cease-fire agreement.

Armistice came into force after 19 months of haggling at 10 am on July 27, 1953. The last four months of fighting alone caused 64,703 UN (mainly Korean) casualties and entailed firing of seven million rounds. This Police Action cost America 157,530 casualties and $67 billion. Total UN casualties were 9,96,937. The Communists suffered a loss of some 1.4 million personnel. The colossal expenditure of shells is depicted in photographs of mounds of empty cases of artillery rounds—77,340 were fired at a hill in two days! 5,88,000 tons of Bombs and over 32,000 tons of Napalm were dropped by the Air Force. The plus point of this War was that it ended the segregation of Negro troops.

An uneasy peace descended. Both sides had taken prisoners. Their exchange started on August 5, 1970. In consequence 183 North Koreans and 5,640 Chinese went back home. The 12,773 UN personnel returned; but 23 Americans, and a solitary Briton, defected, it led to the issue of a 'Code of Conduct' for American servicemen.

India and NNRC—an acute problem arose about the 21,805 North Korean/Chinese POWs who were not willing to be repatriated. The communists claimed that they were being held against their will and were indoctrinated by virulent propaganda and promises of a good life to refuse repatriation. Surprisingly, a very quick agreement was reached between them to refer their cases to a Neutral Nations Repatriation Commission (NNRC).

General K.S. Thimayya of India was called by Prime Minister Nehru on August 18, 1953, and appointed the Chairman of the NNRC. Its four members were from Sweden, Switzerland, Poland and Czechoslovakia. They operated from the Demilitarized Zone (DMZ), which also housed the "Indian Village," where the examination of the North Korean prisoners took place. The Chinese officials, under supervision of Indian officers, interrogated them at great length. There were scenes of violence and long harangues between the prisoners and the communist officers. At times, the prisoners became violent and had to be restrained employing as many as three Indian jawans to control just one of them.

General Thimayya, as the head of the NNRC, was advised by S. Dutt of the ICS, and later by T.N. Kaul of the Ministry of External affairs. Brigadier B.M. Kaul and P.N. Haksar on his staff provided the secretariat. Major General Thorat commanded a 6,000 strong contingent of Indian Army, and Brig Kartar N Dubey acted as the judge against atrocities committed against prisoners.

General Thimayya arrived at Tokyo at the end of September 1953, met General Mark Clark, the Supreme Commander, at his HQ at Tokyo. But his main business was with General William K. Harrison of the U.N in Korea and General Lee Sang-Cho of the NKPA. His mandate was to maintain strict neutrality.

Our troops arrived by ship at Inchon, but as the South Koreans did not want them to travel across their land—they were flown over to the DMZ (Demilitarized zone) in helicopters. The DMZ was a bleak, four-kilometer wide strip of territory, where our troops lived in sort of "Wana huts," concrete block walls, and tented roofs.

The prisoners started arriving, with their flags flying, bands playing and shouting slogans, on September 10. They were moved into camps of 500. They were taken out in-groups of 25; served tea and biscuits and then brought in individually for ascertaining their wish. While being questioned they were allowed to be addressed by their compatriots. This sometimes led to some violently objecting. Such prisoners had to be held down physically by Indian soldiers.

In final analysis, it was not possible for all the 30,000 prisoners to be examined in the limited period of 120 days. The NNRC released the prisoners from its custody on January 23, 1954. Majority of the North Korean prisoners did not want to be repatriated. According to Humphrey Evans, author of "Thimayya of India", four percent of the prisoners chose repatriation.

Brigadier K.N. Dubey with Colonel D.M. Sen, the Judge Advocate General, as his adviser, acted as an International Judge. They travelled by air to Tokyo and then flew in a UN plane to Seoul. He was the President (Judge) of the Court-Martial, to try cases of atrocities committed against the POWs. There were allegedly cases where American prisoners had been shot in the back of their heads with their hands tied behind their backs and left to die.

Dubey acted with great dignity, held court for some four months, dealing with this most difficult problem evenhandedly. I met him to find out, what actually happened, for example if he had to wear Judge's robes. No, he told me—it was very cold and bleak in the neutral zone, he wore his warm uniform and a duffel coat.

To show true neutrality, half his hut was in the North Korean and the other half in the South Korean side of the dividing line, in the four-kilometer wide DMZ. To deal with this most difficult problem evenhandedly required a lot of tact and patience. Language presented the greatest difficulty—there were three sets of translators—one each for the two contestants and the final in support of the Judge. There was constant bickering over the translations. These were managed by replaying the voice recorders. Being neutral meant that all actions, both official and private, had not only been equally preformed and must be transparent so that these also "Appear to be Neutral." This indeed was the most difficult task. Kartar Dubey tried some 12 or 13 cases over a period of three and a half months. Till the Americans got the United Nations to declare the court "*Inchoate*", the legal term to describe that the court was dissolved.

He was very careful and had been on a short break to Japan on the Western side and was due to visit China next, to keep the balance, till events overtook his resolve. In one upmanship, Bijji Kaul and T.N. Kaul took the centre stage on the plea of evenhanded dealings manipulated that they visited China, and lunched with Chou En-lai. General Thorat, Commander of the 6,000 troops of "Custodian Force India" (CFI), performed a most demanding and dangerous task. His troops belonging to our elite regiments—the Jats, Rajputana Rifles, Gurkhas, Dogras and the Kumaonis—provided the backing. They lived in camps in the bleak and bare cold hills of the Neutral Zone. It was a hard life, in the extremely cold winter with little amenities. Both sides of the DMZ had barbed wire fences and watchtowers, which were manned by day and night. Panmunjom, where the armistice had been signed, was the nearest town.

The atmosphere deteriorated till one day General Thorat, a Maratha-of high personal courage, went unarmed into one of the Chinese camps.

The Chinese outraged by the terms of their repatriation, had captured and held hostage an Indian officer. Our troops had surrounded the camp to free their comrade but Thorat told them to hold fire. Accompanied by an escort, a mere handful, he entered the camp. Seemingly oblivious of the tension, he strolled up to the senior Chinese officer; the meeting was in a highly charged atmosphere.

Thorat took out his cigarette case, and offered him a smoke only to find that it was empty. At a loss of what to say he turned round to the Chinese Commandant and said, "What sort of Chinese are you? My men and I have been your guests for an hour and you have not even offered us a cup of tea or a cigarette. Where is the hospitality for which your race is so renowned?"

The Commandant took some time to register the message. The mutinous Chinese stirred menacingly, but after a short while the Commandant barked an order for refreshment.

The preparations for tea got going–the ice was broken, and our officer was released unharmed. CFI stayed on for six months and established their and India's reputation internationally.

The NNRC did not fully succeed in its mission as the communist extended their explanations and arguments for several hours, therefore only 359 POWs could be examined. All the same Thimayya won international acclaim for his masterful handling of the extremely tense situation and was awarded a "Padma Bhushan" in recognition of his role as the head of a United Nations Mission.

Notes

1. Lieutenant General K N Dubey PVSM
2. Lieutenant General B M Kaul

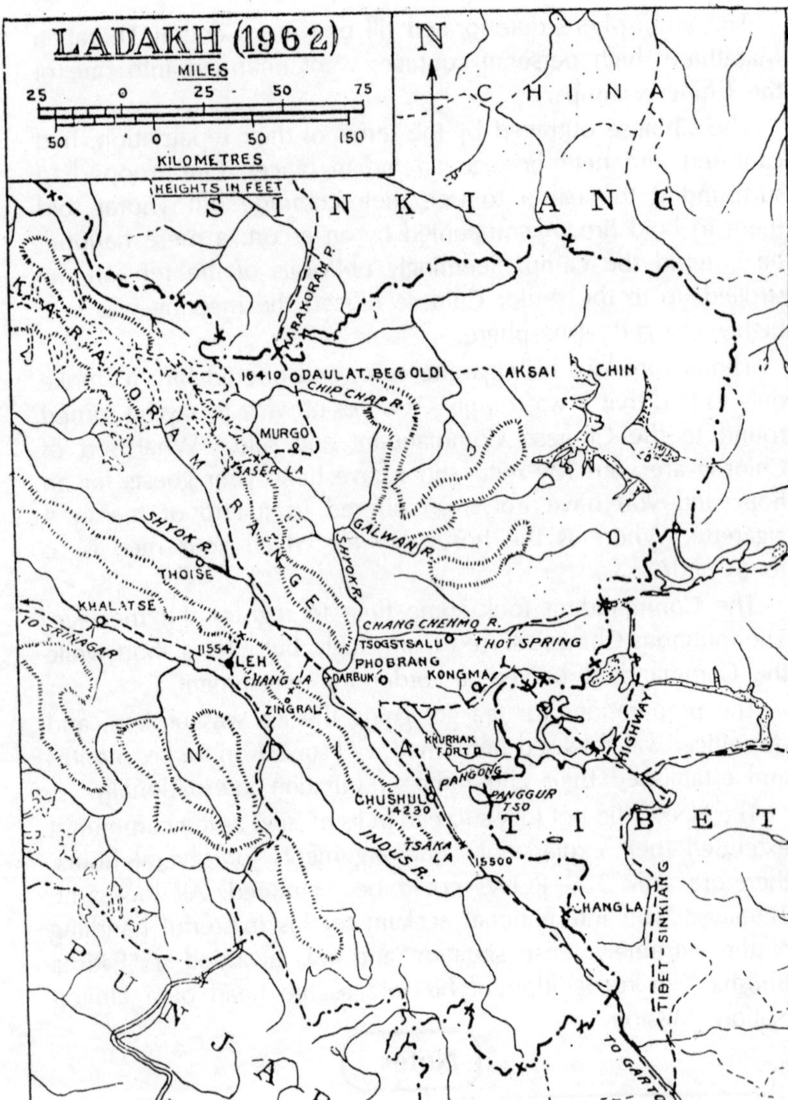

(Boundaries not to scale)

❮9❯

Kaul and the Dragon

The great Chou En-Lai, at the time of Nehru's funeral, called him the greatest "Arrogant" he had known. He certainly got it right. A few days before the Chinese attack, on his way to Sri Lanka, Nehru announced that he had ordered the army to throw out the Chinese. Some years earlier on his way for his first visit to China, he was passing through Rangoon (now Yangon). The Prime Minister of Burma and his cabinet colleagues called on him in the evening. He was having dinner and displayed his boundless arrogance. He made them wait whilst he finished his meal. After it, he waxed eloquent on the neighbourly relations. When asked about India's boundary problems with China, he told them there were none.

Our relations with China started with Nehru's visit to China in 1954. The slogan –"*Hindi Chini Bhai Bhai*"—emerged. But it appears that Panditji was not happy. Unlike in Afghanistan, the MacMahon Line in the East was not marked on the ground. There was a general statement that it shall be on the watershed, which is difficult to demarcate—there is always another hill beyond. As a result of this gnawing doubt, Panditji ordered a reconnaissance of our Northern border, by General Kalwant Singh, the CGS (Chief of the General Staff}, assisted by the Surveyor General. This secret document has not seen by many. Some say it was kept back by the Ministry—till Krishna Menon read it. Brigadier Shanti Malik, the Military Attaché in China, reported the construction of the Aksai Chin Road, by the Chinese in our territory. In 1956, Capt Iyenger's popularly called 'Kasim' party, to verify the extent of the encroachment, was captured, warned, and sent back by the Chinese. He was awarded an Ashoka Chakra. Later, when they killed a police party of eight men under Havildar Karam Singh in 1958, some reservations arose in Nehru's mind.

After Chou En-lai's visit to India in 1960, Nehru visited our forward areas in NEFA. He was accompanied by Lieutenant General K.N. Dubey, who having been warned, not to take him to heights, took him as far as the base of the Sela Pass. On the way he visited several Tibetan refugee camps, distributed sweets to their children, and spent a night at Bomdila. He expressed his doubts to Kartar—our policy took shape. Possession being nine points of the law, our troops were told to occupy forward posts. These were within our territory as can be seen from the excellent maps in Hoffman's book *India and the China Crisis*. Lack of roads in the area however caused problems in the maintenance of the posts.

Tracks were ordered on the local PWDs who, complained that the Army was not paying. The Cabinet Secretary, Mr Bhagwan Sahai, set up a Border Roads Organisation. General Kartar Dubey, the first Director General of Border Roads, reopened our relations with Bhutan. Brigade Jeganathan tasked to build the eastern Bhutan road, built it with remarkable speed—it was through by May 12, 1962. Later it enabled some 3,000 officers and men from Sela and Tawang to get home after the attack by the Chinese.

The Chinese objected to our forward posts. Hostilities broke out first in the NEFA on October 20, 1962. The one-month war ended with a unilateral ceasefire by the Chinese on November 21. No way can it be called the greatest disaster to "Indian Arms." Only a small part of our Army, stretched over a vast undeveloped hilly region without logistic support was involved for one month.

The Government ordered the army to evict the Chinese without the help of fighter bombers, it was afraid of Calcutta being bombed. The greatest casualty was Nehru's confidence. Having misjudged the threat, he let the war be fought by a General from his sick bed, a thousand miles away from the battle. At the same time, he appealed for help from the USA, UK, Canada and Australia. A strange admixture with little understanding of military, he again disobeyed the Duke of Wellington's maxim: *"A great country can have no such thing as a small war."*

Today, 40 years after the event, all documents, including General Henderson Brook's report, are secret. A perilous state of affairs, as no military organisation can progress and improve its working—without introspection.

Bijji Kaul was a go getter and forceful person but the Chinese, well supplied, operating from higher ground were basically at an advantage against us. Our troops were cold, hungry, and lacking even ammunition. As the CGS, it was Kaul's duty to arrange for the requisite weapons and improve supplies to enhance their defence capability.

The war started in the North East along the undefined mountainous border in NEFA, where our troops held the near bank of a hill stream, with some log bridges across it. The position was dominated by the Chinese who held the hill above called Thagla Ridge. Towang and Sela Mountain positions were our back up defensive positions. Our reverses were blown out of all proportions. Books have been written by Bijji Kaul, Niranjan Parshad the Divisional Commander, Monty Palit the DMO and Dalvi 7 Brigade Commander. The story emerges on sifting through the background information. Knowing them all, I find they have tried to justify their actions in retrospect. Operations took place in three sectors. Ladakh in the northwest, Walong in the extreme northeast and Towang-Sela in the centre. Only a Division worth of troops was involved in the operations in Ladakh and central sector of Towang, which was in the limelight; Walong on a limb had only one Brigade.

The Towang sector under General Kaul, who hungered for publicity, stole the show. At one time it was swamped by 21 visiting Generals who hogged the scarce transport, needed for the troops fighting at mountain heights, in summer uniforms, short of boots and ammunition. I remember my wife receiving a letter from Joyce Wright from England, "Our Jawans fighting in the snows without woolen clothes, in tennis shoes, please deny it at once". Colonel Thomas in his book, published in Australia wrote, "The Battalion left its winter clothing at Tezpur, like Christmas is coming?" Brigadier Dalvi stated, "He could not sleep very well due to cold and the troops were having their tea with salt instead of sugar."Our troops armed with old bolt operated rifles, without proper artillery support, were even denied fire support from the Air Force. The Government was afraid that if we used fighter bombers, Calcutta might be bombed. The Joint Secretary issued a written order to "Throw out the Chinese" which the Chinese were in no mood to accept!

Having personally known the major actors and sifted through a lot of information my true story is spun round the two Generals in command in the Ladakh and NEFA. Both had been my friends since our days at the Government College, Lahore. Daulat Singh joined the Army first and as the senior, commanded the troops that held the Chinese in Ladakh. Kaul, who was the Chief of the General Staff (CGS), the key job for directing operations, was instead literally pushed into the battle in the Eastern Sector to gain the glory of "Throwing out the Chinese just because he was a distant kinsman of Nehru."

Portents of War

Mrs. Prem Bhatia told me of how at her ladies coffee party, in the September of the fateful year, her Egyptian guest accidentally spilt her coffee. Embarrassed, she offered to 'read' the cup. Suddenly, she turned very grave saying, "I see War" and added, "It is in the East." May be it was the tip off to Prem, who was first with the news of Bijji Kaul flying to Tezpur to "Evict the Chinese." Yet, surprisingly on September 8[1], when the Director of Military Operations (DMO) received the report of Chinese surrounding our Dhola post in the NEFA, Palit, and the DMO chose to go on a cruise with the Navy. General Thapar, the Army chief, did not bother to recall the CGS, even when on the September 22, Harish Sarin, Joint Secretary, issued to the army the order to throw out the Chinese. Sarin acted under the authority of Raghuramiah, the Deputy Defence Minister, not even a member of the Cabinet. The country declared a virtual war, while both Prime Minister Nehru and Defence Minister Menon were out of the country, they did not even consider it necessary to get back.

The Army Chief issued a perfunctory warning signal to Commands. Ten days later, when Prime Minister Nehru returned from abroad, Kaul came to Delhi to meet Panditji. It was on October 2, Gandhi's birthday, a holiday. He resumed the appointment of CGS and in a great hurry by late evening of October 3, he was appointed as the Corps Commander of a nonexistent IV Corps, to take over from General Umrao Singh's XXXIII Corps which had been in position to oversee the work of a Division plus a Brigade worth of troops in the NEFA. This changing of horses' midstream on the eve of operations will always be a question mark. Was the object to glorify Kaul's ego or pacify Nehru?

General Daulat Singh, a tall and handsome cavalryman, he cut an excellent figure on the Polo field. Having served in horsed cavalry, he too lacked experience in mountain warfare. Having studied at the Imperial Defence College, he gained insight of higher command. He performed better, as he had less interference. He handled the vast Ladakh front, an operational area of more than 400 kilometers from Daulat Beg Oldi in the north to Pangong Lake near Chusul airfield in the south. His troops made orderly withdrawal, and inflicted many casualties on Chinese.

The Chinese were well supplied from their bases, by the Aksai Chin road. The area, a desolate plateau called the Soda Plains or Desert of White Stones. Their three divisions vastly outnumbered our forces. The Government had not agreed to Daulat's request for additional troops put up in 1961, only two battalions, one J&K Militia and one regular battalion were allotted to him. Skirmishes in the Ladakh area had been going on since long.

The Chinese first threatened our Chip Chap post in May—our men held fast—and the Chinese withdrew. Galwan post had a similar experience in July. It was again attacked on the night of the October 19/20, a detachment of 372 Field Company headed by a JCO, was wiped out. The Chinese had a superiority of five to one. The next day they concentrated heavy mortar and artillery fire against Sirjap post, along the northern shores of Pangong Tso Lake. The Gurkhas fought heroically to the last man, but the post was overwhelmed. Some Sappers of 9 Field Company were taken prisoners. The fall of Sirjap post and threat to Yula resulted in a fallback across the Pangong Lake. This lake, because of its enormous size and strong winds, had huge waves making it tricky to navigate. Yet, our Sappers used two storm boats to ferry back the survivors under the cover of darkness. One of these boats was hit and sunk by enemy fire. The night was pitch black, the lake rough and the enemy was firing volleys of machine-guns across it. Despite these hazards, Lance Naik Raghavan, a hundred yards ahead on the other boat, turned it round to rescue the survivors in the boat. He brought back three and kept on trying till a fast Chinese Tug with a searchlight chased him. He was awarded the Vir Chakra.

In the North, Daulat Beg Oldi was threatened on October 22. It had a slender line of supply based on an indifferent airstrip. It was not possible to hold it, so at about 5.30 pm, it was decided to withdraw. Lieutenant Karir[2], the OC of the Sapper detachment, wrote, "I decided to stay and help my bunker mate in the evacuation. Before we left everything, which could be of use, was destroyed—torched. We were expecting to be fired on during the withdrawal by the Chinese, but for some mysterious reason they did not. Possibly they were deceived by a decoy signal sent by the post Commander—asking permission to stay on after he had been given permission to abandon! We marched over a hundred miles in the next eight days. The first few miles were along the frozen bed of River Chip Chap. The lorries and jeeps were driven along the river bed, but had to be abandoned one by one as the ice cracked or the boulders obstructed the way. A patrol of three or four Chinese was seen following our movements from a ridge about 2.000 feet above us. The first night halt was in a bend of the Chip Chap, with a glacier right opposite us. Even eight blankets proved inadequate for the biting cold of the night. Every fifteen minutes, I kept on burning pages of a book I found in the pocket of my Parkha trousers. People with sleeping bags were better off. In the subsequent days, we walked through majestic countryside, utterly lifeless and arid, with not even a blade of grass for miles and miles. We trudged for eight hours along the glacier of Saser La, even as it snowed. Some Jawans cut off their well-tended moustaches because the moisture of their breath kept freezing on them. IAF Helicopters dropped us food packets. Puris and potato curry, (hard and frozen) tasted better than almonds, after days of subsisting on *channa*. Drops of ready to eat food packets were by the courtesy of Sardar Partap Singh Kairon, the Chief Minister of the Punjab. He had appealed to the housewives to make parathas for the Jawans and made arrangements for collection and supply of thousands of stuffed parathas to various posts.

The post at Demchok was captured by the enemy on October 27. There was no alternative for our Jawans but to make an orderly fighting withdrawal which they did—they withdrew bravely—and steadfastly—a tribute to the good command and control on this front. Their orderly behaviour was of great morale value, and testified to the leadership, than can be shown by writing any amount of rhetoric. In his book "India's China War", Neville Maxwell praises the conduct of the war by General Daulat Singh—"He was methodically and rapidly building up strength on the threatened Chinese front by pulling troops out of Kashmir. All the Command's transport reserves were pooled. By the first week of November he established a Divisional HQ under General Budh Singh MC and Bar and also got together an additional Brigade of four Infantry Battalions. By Nov. 17, another Brigade had been inducted to Leh. On the Eastern front, however, there was no such decisiveness or dispatch."

On the move of additional Brigade to Leh, the road to Chusul was made fit for three tonners. 26 Divisional Engineers were brought from Nagrota for extension of the Chusul and Fukche Airfields and subsequently airlifted back to Chusul in October to prepare defences, operate ferries to Sirijap and Yula posts across the Pangong Tso. The fighting continued in the Pangong area, as well as in the Spangur lake area. Defence of Chusul airfield surfaced with PSP (Pierced Steel Planking) withstood the heavy traffic. It was declared forward vital ground, and defended by Brigadier T.N. Raina's 114 Brigade. The airfield was held by 1/8 Gurkhas on the left, with 13 Kumaons on the right, the centre was covered by the old faithful—AMX tanks which had been dismantled for transit and reassembled at Chusul. When the enemy supported with elements of armour attacked Chusul, it encountered anti-tank mines which had been laid.

The gallantry of our troops displayed was of the highest order. I can do no better than quote from Major General Jagjit Singh's book, "Ladakh", published by Vanity Books, Delhi. The Rezen La feature 5,180 metres high, was held by C Company of the 13 Kumaon Regiment, under Major Shaitan Singh. The position was well dug in, but mines had not been laid as these were not available. The main weakness was that due to shortage of Artillery, the fire support was limited to Mortars. We had no guns available for counter battery fire. Yet, the position was held unmindful of losses till the cease-fire, in the face of Chinese rockets and flame throwers. Out of total of 118 all ranks, one officer, two JCOs and 106 other ranks laid down their lives. One JCO and four other ranks managed to return alive from the ghastly hell at Rezang La on November 18. One burly Ahir lifted a Chinese and threw him down the Rezang La. Even the Chinese, recognizing the bravery of our men covered the fallen with blankets, and stuck their bayonets in the ground.

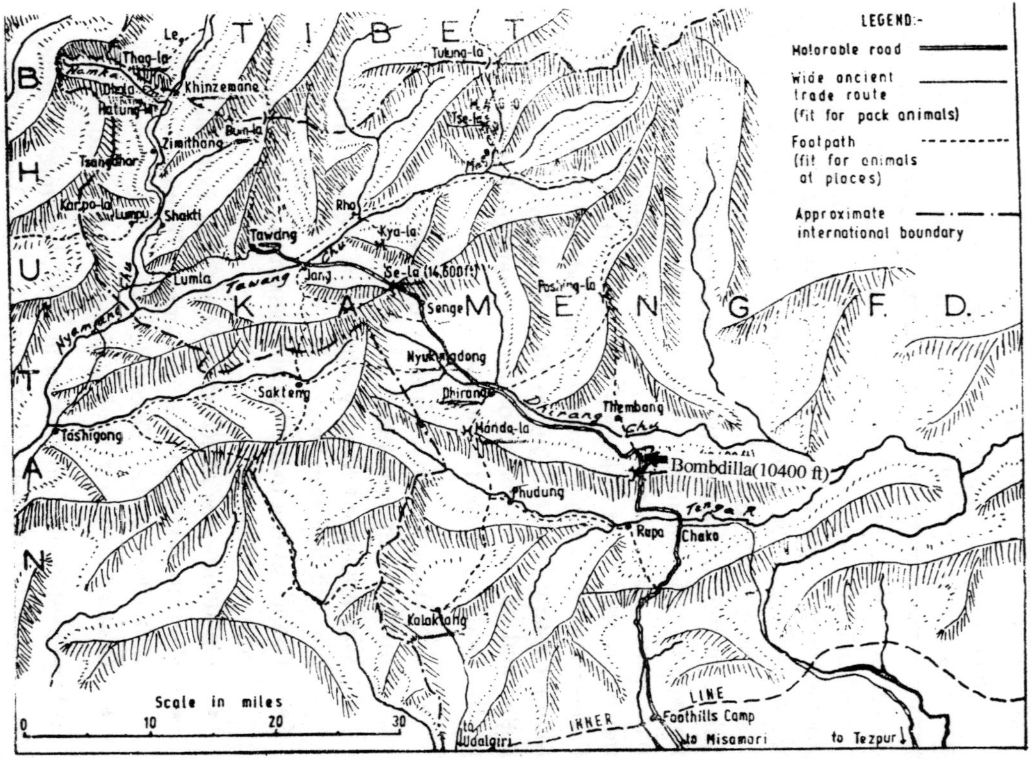

Kameng Frontier Division (Boundaries not authenticated)

NEFA—Kameng Division

Kaul of medium build, with a pleasant open countenance, was very active both physically and mentally. He spoke with confidence. He had a marked sense of humour, and was always game to play practical jokes. After only a two and a half years service in the Infantry-he opted for the Army Supply Corps (ASC). To get over the stigma of having spent the War in the ASC and 'Fauji Dilkush Sabha', he got himself a much sought after appointment as the military attache in the USA. Bijji could be depended upon for grabbing opportunities. He wrote in his book "The Untold Story" how when Pundit Nehru went up to Shimla in the summer of 1945, to meet the Viceroy, he lined up with his wife Dhanno, a distant relation of the Nehru's. Met him again and thereafter became the "Chosen One." Nehru moved to York Road residence on Sept 7, 1946, as the virtual Prime Minister. Auchinleck took the hint and wrote to the Defence Minister asking for B.M. Kaul as his Liaison Officer. Kaul aimed higher and taking advantage of the new relationship; he moved up the ladder described as follows:-

- Breakfasted with Nehru to discuss General Sir Roy Butcher's appointment.
- Was appointed Military Attaché in the US, ordered to buy Mitchell Bombers.
- Posted Advisor to the Indian Mission to the UN—lunched with Gromyko.
- On return was attached to Ministry of External Affairs.
- Went to Korea on staff of General Thimayya—lunched with Chou En-Lai, December 22.
- Detailed to arrest Sheikh Abdullah—August 8, 1953.

When Pakistan attacked Kashmir in 1947, he wrote to Panditji, from the US that he wanted to return to fight. Was posted to Kashmir state, J and K Rifles for five months. Soon afterwards, he was given command of 11 Infantry Brigade. From there, he became Director Territorial Army, and then commanded the famous 4 Division. As the Divisional Commander, he misemployed his formation on a well-publicised housing project for the troops. It presumably qualified him to become the Quarter Master General and take over unprecedented control of the Corps of Engineers. Raghuramiah, the Deputy Defence Minister, on receiving this proposal refused to examine me and submitted a thumping aye.

Finally, he was catapulted as Chief of the General Staff and made a Corps Commander, by a complacent Thapar who wanted to back the rising stars. Books like "After Nehru What?" helped to weave the picture that he was destined to reach the top. To eliminate a competent competitor, he instigated an inquiry against Major General Sam Manekshaw but General Daulat Singh, the Inquiry officer, found no substance in it. Kaul, in his book *"Confrontation with Pakistan"* says that the first inquiry against Sam was by the Intelligence Bureau.

We had been together at Staff College, Quetta, in 1942. Our paths did not cross for 16 years till I was CGDP with Krishna Menon. He called on us to get an introduction to him. An essential rung in the ladder. He knew Timmy, the Chief, having served with him in Korea, and had access to Panditji. He did not want any impediments. There was a misreading regarding his rapport with Krishna Menon—both were dependent on Panditji.

General Bogey Sen, the Eastern Army Commander, had ordered General Umrao Singh, commanding XXXIII Corps to send a Company to Tsangle post. Umrao rightly objected to Eastern Command issuing such detailed instructions. The rift between the Corps and the Army Commander soon came into the open.

Thus, was dealt a despicable under hand way to boost the image of Kaul—(still on leave). A proposal was put up for raising a new IV Corps, to take over the conduct of operations in the NEFA area from XXXIII Corps commanded by General Umrao Singh. This was however a weird solution to the situation.

On his return to Delhi, Bijji Kaul was appointed to command IV Corps. He in a clandestine manner selected his staff overnight, at a meeting at his residence, at which the Assistant Military Secretary was present. To issue formal orders as dictated. Between midnight and 2 a.m. on August 4, they were to fly out to their doom to Tezpur. To enable him to return to Delhi subsequently as CGS, the post was kept vacant: Major General J.S. Dhillon of Bengal Sappers was made officiating CGS. Never mind that it was a key appointment at a crucial period. It was a ridiculous situation.

Kaul's meeting ended in the early hours of that morning. An officer in whom I have implicit faith told me of the strange portents of a natural calamity. They suddenly heard a loud crashing noise—rushed out to look, and found that a very large and tall tree had fallen across the entrance gate of the bungalow. This was beyond their understanding as there was but a very gentle morning breeze—it seemed strange was it a warning? Palit the DMO, back from his naval cruise, met Kaul as he was departing for the airport.

He met Prem Bhatia of the *Times of India* who was with Bijji. In the morning, the paper published the story of the gallant new Corps Commander flying out to 'throw out' the Chinese! Bijji landed at Tezpur in the afternoon of the October 4, with his staff. Surprisingly, he did not accept the helping hand offered by his friend, General Umrao Singh.

Enthusiastically, he set off to the front, met Major General Nirinjan Parshad, the Divisional Commander, on the way. Both of them reached the front on the mountain stream. It was a tough journey through mud and sleet, he had to be carried on the back of a porter! He saw the enemy engaging at our, our men clad in summer uniforms holding the indefensible positions on the mountain stream known as Nam Ka Chu. The troops were even short of ammunition. The Battalion had only two Medium Machine Guns (MMGs). Later I met a young officer of the Rajputs the Battalion involved. He related to me how the Corps Commander discussed minor tactics. He asked for volunteers to take on an enemy picket, instead of dealing with logistics, sector defence, and larger issues. On the morning of October 10, the observation post where he stood came under fire by the Chinese automatic weapon. That unnerved him.

He returned post haste to Delhi within a week. On October 11, he had a conference with the PM in which he highlighted the lack of supplies but forgot to mention the most important thing that is the lack of automatic firing weapons and shortage of ammunition.

He returned to Tezpur on October 13, and kept up his busy schedule of visits. Later, on October 16, he sent a dispiriting signal, recommending a pull back from Tasangale. The Chief, General Thapar did not agree, and ordered Kaul to be more aggressive, though Thapar too lacked battle experience.

Next day, not used to the heights, Bijji fell seriously ill, and had to be taken to his house at Delhi. Hard to believe but true, he was allowed to run the Corps, from his sick bed, some thousand miles away. When Palit suggested that General Sen move up protem to take charge of the battle, General Thapar told him, "General Kaul will continue in command." During the time, he lay sick in his house at 5 York Road, Delhi, he was visited by dignitaries from the Prime Minister down.

He advised that besides raising additional forces within our resources we should seek military aid from some foreign power. After the fall of Towang on October 26, Krishna Menon called and asked Bijji to reduce his suggestions to writing. Kaul after consultations dictated a short paper, which the Cabinet Secretary himself collected. In response to our appeal, America, the U.K, Canada and Australia responded positively. As recorded by Bijji, he was kept informed of the situation by phone. He was allowed to fly back on October 29. General Nirinjan Prasad, met him and represented that Towang (7 Brigade) had been evacuated on October 23—in consultation with Bogey Sen, the Army Commander. Bijji, after resuming command, ignored the instructions of the Ministry of Defence, allowed newspaper correspondents including Edward Behr of the *Time* magazine to visit and report on the operations—consequently received some flattering publicity. Highly ambitious, he liked to be known in the World.

There was no end to visitors—the Chief of the Army Staff, Thapar and the Army Commander Sen were with him on November 7, and again on November 17. On November 17, Sen rang up Nirinjan and conveyed a curt message, relieving him of command of 4 Division. Minutes after two helicopters landed, the first with Major General A.S. Pathania MC who was to take over from Nirinjan, the second brought Lieutenant

General Harbakash Singh, the new Corps Commander. Accompanying them was Major General Mohinder Pathania—who arrived in advance of his 2 Division HQ. Army HQ appeared to be bent on justifying the importance of Kaul's Corps controlling two divisions, the 4th in the central zone and a new 2 Division HQ was to supervise 11 Infantry Brigade at Walong in the east.

General Palit, the DMO, who knew the area, also, came. They talked in whispers to each other. Next day, they went forward to 'look see'. Harbaksh the new Corps Commander called a conference of heads of staff. When they were having a drink in the mess, they heard a radio announcement that General Kaul is coming back to resume command. Harbaksh was at a loss for words—a scene more suitable for a hilarious comedy. He retired to a tent. Next morning, he was flown to Siliguri to establish his XXXIII Corps HQ. There, in response to the request for help, C -130 transport planes of the US Airforce started landing on November 3, bringing troops from the Punjab, automatic rifles and winter clothing from the States.

Walong, at the north-east end of NEFA, and near the tri-junction of boundaries of Tibet-Burma and India, was in very capable hands of Brigadier Naveen Rawlley MC, a WW II veteran, commanding 11 Infantry Brigade. Kaul visited Walong on the November 12, and again on November 15th accompanied by General Dhillion on tour from Delhi. While Dhillon went back, Bijji spent the night at HQ 2 Division. He asked the Air Force to fly him near Walong as he wanted to spot Brigadier Rawlley, in the withdrawal. He got back to his HQ at 1930 hours. General Thapar told him that Pathania 4 Division Commander was trying to contact him. A peculiar situation the Army Commander and the Chief of Staff visited Kaul's HQ; he was not present. Normal military custom was thrown to the winds. He apparently spoke to Pathania and wrote a long rigmarole note, "Delegate to you the authority when position is untenable." He failed to become a great Captain as he was not even present at the battle.

The worst happened on the evening of November 18, our famous 4 Indian Division, left its dug in position at Sela Pass under Brigadier Hoshiar Singh, Legion of Fresh Honour was ordered to pull back inspite of Brigade Commander's serious objections. No clear orders appear to have been given. We lost credibility at Towang-Sela. The basic cause was, wholesale change of unit and formation Commanders. Divisional and three Brigade Commanders were replaced, just before the battle which never took place.

The Indian army paid heavy price for the abject failure of its political masters and the Army General Command. The total number of casualties in the Indo-China war of 1962 in the Eastern and Western Front were:

Killed	:	1,383
Wounded	:	1,696
Prisoners of War	:	3,968 (of whom 26 died in captivity)

The Aftermath

T.T. Krishnamachari, generally referred to as TTK, took over as the Minister of Defence Co-ordination. I was then in a civil employ at Calcutta. On receipt of a message from P.V.R. Rao, I caught the first train to Delhi. Calcutta Mail took almost 36 hours. As I got into the taxi at the station the driver told me that the "Chinese have run away." It was November 21. The credit for rearming the Army must go to TTK. The Ishapore Rifle was finalised by General Williams and put into production. Our defence potential was strengthened by the addition of three factories—a lasting monument to Rear Admiral Daya Shankar.

Finally, the story of 186 officers and over 3,000 valiant men who got back to India from Towang and Sela as told by Lieutenant Colonel Thomas of 4 Grenadiers, whose men, plus 1/9 Gurkhas and the Punjabis, snaked their way on unknown tracks for a fortnight. Ahluwalia, of the 1/9, badly wounded requested to be left in the shelter of a rock, he did not want the others to be held up. Cold and hungry they spent the nights on freezing heights on windy glaciers—harassed by the Chinese. They finally sighted the border post of Tashingang Dozong in Bhutan. Here they were welcomed to an Army camp organised by Brigadier Jeganathan who had put his men on half rations to feed his guests. Not only was there food and shelter, there were helping hands. There were trucks to give them rides on a metaled road back to India. Helicopters to fly back the seriously wounded to the hospital at Tezpur. The crowning glory was when they were visited by our great scholar President Radhakrishnan.

Balwant Aluwalia, whose wife Bai was heartbroken, was taken prisoner by the Chinese finally, came back. General Tewari, then a Lieutenant Colonel of the Signals, also taken prisoner, has written an interesting account of their treatment. Before being repatriated, senior officers were taken on a grand tour, from Lhasa to Beijing—a sign of their cultural heritage to show that they are an enlightened people and allowed the prisoners to recieve Red Cross parcels, to keep in touch with their families.

Finally a little known book by General Tewari of the Signals *"A Soldier's Voyage of Discovery"*, describes the treatment of our prisoners by the Chinese. Naturally, they tried to brain-wash them—gave them long lectures. But their treatment was by no means inhuman, prisoners received Red Cross parcels, were allowed to write home; my friend Balwant Aluwalia even received letters and telegrams. Lieutenant Colonel Rattan Singh, who had previously been a prisoner in Malaya, tells of the terrible time he had, at the hands of the Chinese as he didn't join the INA.

Wounded and prisoners were exchanged quickly to prevent souring of relations and to dispel any thoughts of mutual hatred. Let me tell you of an incident related to me by Late General M M Khanna. During the operations in the Central sector—a company runner of the Kumaon Regiment while taking a message, jumped into a slit trench for a smoke. It was dusk—he couldn't see who was sharing the trench but asked *"Sathi bidi hai"*. The Chinese at the other end replied *"Hindi Chinee Bhai Bhai"*. Realising his error he quickly jumped out and was on his way without let or hindrance!

Notes

1. D.K. Palit's book
2. Lt Col O.N.Karir

1965-Indo-Pak War

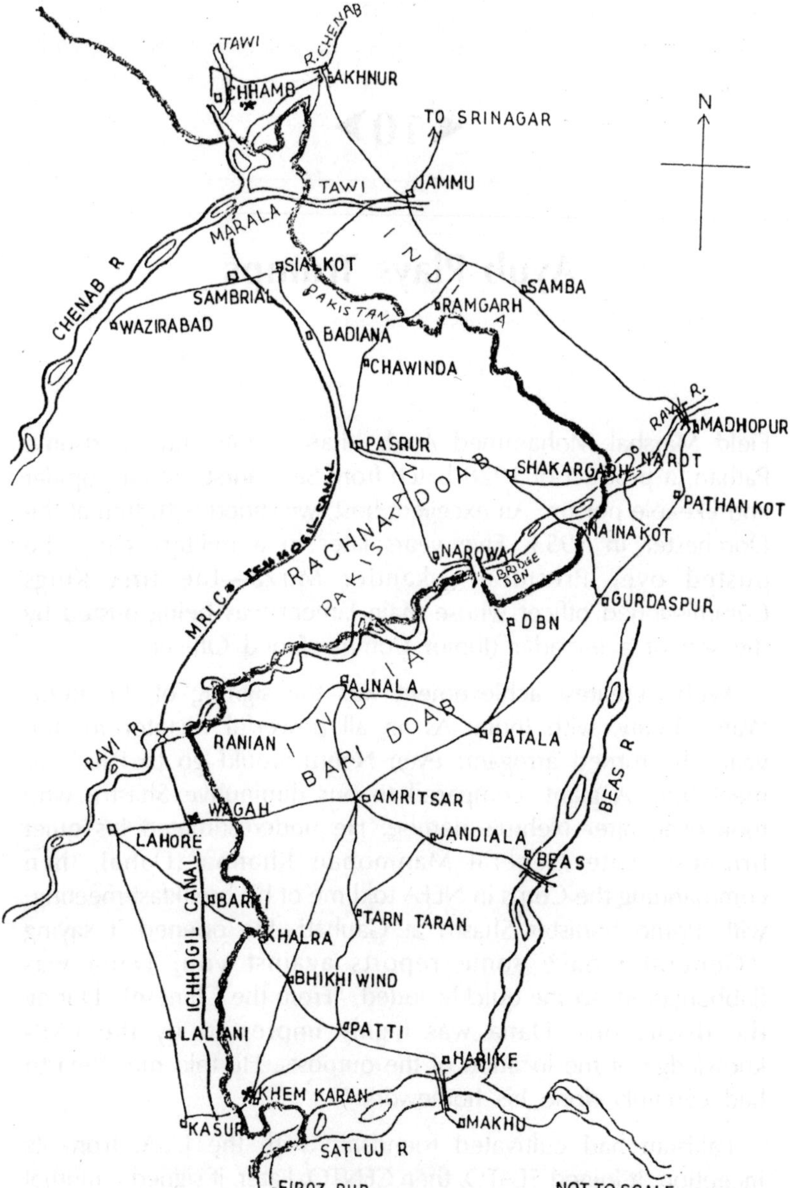

❮10❯

Ayub Plays Trumps

Field Marshal Mohammed Ayub Khan, a tall and handsome Pathan, a product of 1928 batch from Sandhurst, was a popular and likeable person. An excellent host, we dined with him at the Dorchester in 1953. Five years later in a military coup, he ousted over President Iskander Mirza—the first Kings Commissioned officer whose main lament was being ousted by the son of a Jemadar (Junior Commissioned Officer).

Ayub's greatest achievement was the signing of the Indus Water Treaty, with India. As an all powerful dictator for ten years, he turned arrogant; even Nehru would go to Palam to meet him. A giant, compared to our diminutive Shastri, who took over, after Nehru's demise. He underestimated his quiet firmness. Late General Manmohan Khanna[1] (Dana), then commanding the Corps in NEFA told me of his breakfast meeting, with Prime Minister Shastri at Gauhati. He opened it saying ,"General I have some reports against you. Dana was flabbergasted, so she quickly added, "From the Chinese!" During the discussions, Dana was highly impressed by the PM's knowledge of the location of the outposts. He told me, "Shastri had certainly done his homework."

Pakistan had cultivated friendship with the USA, from its inception. It joined SEATO, then CENTO. Later, it signed a mutual aid pact in 1954 and got 2.5 billion dollars worth of aid mainly in military hardware.

In 1959, Ayub played his trump card. He signed a ten-year agreement to provide America a base at Peshawar for their U2 spy planes and provided facilities for setting up electronic listening posts for surveillance of Russian missiles at Badaber. This opened flood gates of American weapons and equipment. A Military Aid Group, USMAG, arrived to equip and train additional 40,000 Pakistani troops.

Hordes of officers went to America to learn American doctrine. The result was that, during 1965 war, we could hear Pakistan infantry, not moving forward and chanting, "Ali-Ali," waiting for armour to lead them into an attack.

A seasoned soldier, Ayub, started with reconnaissances. In 1963, he went round the coastline past the Rann of Kutch aboard cruiser "*Babur*". which led to an encounter with our aircraft carrier "*Vikrant*" of Cochin. Rear Admiral Satyindra Singh described it:-"The fleet was resting after a week's intensive flying, when early in the morning whilst Admiral Krishnan was at puja—the signals officer rushed in to report "A largish echo which is moving too fast, for a merchant ship, and is heading towards us, range 20 miles and closing in fast." Alarm for action station was sounded. Steam was raised, cable party stood by to slip anchor. Krishnan ordered two aircraft to be in readiness. The silhouette was recognised as that of Pakistani cruiser *Babur*. Soon it was in visual range. Krishnan knew Syed Mohd. Ehsan, its Commander. He signaled, "Syed do not come closer." The reply came, "Krish have Ayub on board bound for Colombo. Thought we will have a *dekho* (look) at my old country. Cordial greetings" and turned away!

Besides improving relations with the Middle East, Ayub befriended Indonesia, it offered to attack the Andamans. Cordial relations with China, which had informed him of the ceasefire in 1962 at least 48 hrs before the event, were cemented. In March 1963, Pakistan gifted away the valley of river Shaksgam or Muztagh a tributary of the Yakand-some 5180 square kms, of Indian territory to China over which it had no right, but covered it by calling it, "Area, the defence of which is under its actual control of Pakistan."

By 1965, Pakistan having acquired 200 Sherman and 400 M-47/48 Patton tanks, secretly raised an additional armoured division, named it the 6th. There were problems in introducing the 60 ton Pattons. Lieutenant General Gul Hassan, tells the story of solution of restrictions to their movement on Pakistan railways. Gul as ADC to Field Marshal Slim had learnt, when faced with a technical problem send for a Sapper. An unsung Major Ahmed Saeed, was given the task. He spent two months visiting railway establishments and finally at the meeting with the Chief Engineer of the railways, in his twenty minutes demolished all his objections. The Pakistan Railways wanted to refurbish the whole of their system, at the cost of the Army!

A tank man, he wrote, "The Pattons looked far more menacing when in our tank parks, than on exercises where breakdowns were not infrequent. The issue of spares and practice ammunition was niggardly." Finally command of the sophisticated Pattons, capable of operating at night, was in the wrong hands. The first Commander of the 1 Armed Division was Major General Haq Nawaz, whom I knew at HQ XIV Army. Gul Hassan's picture of him as one devoid of technical skills with keen knowledge of the small oil bottle in the butt of the 303 rifle—fits him to the 'T'.

Pakistan had also bought the notorious F 104-A Star fighters, *"widow makers."* The Germans had found them too dangerous to fly, and were glad to sell them off, with a nod from America. These missiles firing Sabre Jets, were however much superior to our Gnats with 30 mm Aden guns. With new tanks and fighters, Pakistan forces were raring to take Kashmir by the force of Arms. Propaganda was unleashed saying how one Pakistani soldier was equal to six Indians. The balance of power was in Pakistan's favour though India had the bigger Army in numbers.

Comparative strengths of the two opposing armies: 1962

	West Pakistan	**India-Western Army**
Armour		
Main Battle Tank	352 (M47 and M48 Patton)	192 (Centurion)
Medium Battle Tank	308 (M4 Sherman)	366 (M3, M4 and M5 Shermans)
Light Battle Tank	96 (M24 Chaffee)	96 (AMX 13) and 48 (PT 76)

After its 1962 war with China, India had raised Nine Mountain Divisions. Towards the end of the war with China; the U.S, supplied much needed winter warfare equipment, gave India aid worth some 60 million dollars. Later, it helped us to build four armament factories.

The year 1964 was a difficult year for India. Events, which piled up including the reported theft of the sacred relic—Prophets' hair in Kashmir. The release and subsequent actions of Sheikh Abdullah, Chou En lai's declaration of support to Pakistan and the death of Prime Minister Nehru. These eventually led to the end of peace.

Knowing that there is no second chance in war, Ayub wanted a quick and decisive war, and had trained a "One Shot Army." Fighting with imported armaments, against an adversary, with its strength based on indigenous production. He did not want to wait any longer. He was afraid that he could run short. That's is exactly what happened. Read on—

This war took place in three time frames. In April 1965, Pakistan attacked in the Rann of Kutch. In August, launched operation "Gibraltar", infiltrated guerillas into Kashmir to take over the state, and when that failed, it finally launched a "*Jehad*" -described as "striving for justice and truth and not a war for territorial ambition", in September.

On April 9, 1965, a Pakistan Brigade from Malir Cantonment near Karachi, captured a small Indian frontier Police chowki, Sardar Post, and followed it up with a larger attack by a Division using Patton tanks against five posts including Chad Bet. I remember Chad Bet (a bet is high ground) when in 1956, this post was retaken by the great soldier General Thimayya. He had mounted such a lighting attack that the Pakistanis departed in haste, leaving their cooking pots still warm! A well boring unit, sent to dig for water, found oil instead. But the Government did not want to spend money on development of roads. We could not move our tanks as the ground on our side was not suitable, being wet land to face Pakistan's 8 Division, operating under Ayub's personal directions.

On April 26, the battalion based at Bhuj had to withdraw; we had no tanks and lacked even anti-tank weapons and mines. It caused great excitement in Pakistan. Sporadic fighting however continued, there was danger of escalation. B.K. Nehru, our then Ambassador to the USA, recounts how he flew to Ottawa to meet Shastriji, and inquire, as to when the skirmishes on Indo-Pak border would abate. The PM told him, "A nation which allows its territory to be nibbled away piece by piece, has no right to exist. When Pakistan vacates Indian territory. " This forthright comment was no doubt conveyed to Pakistan by the British PM—Harold Wilson. Pakistan withdrew. A cease fire, was then arranged effective June 30, with an agreement to refer the case for arbitration by the World Court. Ayub and Shastri shook hands, our troops went back to their barracks.

Those days, Sir Mohd. Zaffarullah was a powerful member of the World Court. To show our neutrality we chose a Yugoslav to represent us. Pakistan chose an Iranian and claimed an additional 30,000 square kilometers, with the boundary at 24 degrees latitude on the basis that the Maharaja of Kutch did not exercise control over it. For years the Maharaja had been abroad. I knew him, as a councillor in London. In his absence, his retainers failed to produce the records. We lost 3,200 square kilometers of our territory. Today, Pakistani maps show Chad Bet as theirs. A note in small print says, "The International boundary between Pakistan and India runs along the middle of the great Rann. This has not been marked and is under dispute."

Operation "Gibraltar'—Five task forces of insurgents infiltrated, to foment local uprisings, in the valley; it was timed for August 8, the date of Sheikh Sahib's previous arrest. But the expected Kashmiri uprising failed miserably. Guerillas who had infiltrated a day earlier to mingle with the locals and attack specific targets, failed to get off the ground. General Akbar Hussain Malik's, (GOC 12 Pak Division at Murree), infiltration behind the lines in Titwal, Tangdhar area was a damp squib. The anti-climax came next day when four captured Pakistan soldiers, spoke on the AIR's 10 pm broadcast. General Musa, the Pakistan C in C, in his book accepted his folly.

To prevent infiltration, our troops captured outposts overlooking Kargil on the road to Leh and Haji Pir Pass, key to movements between Punch and Uri. Story of its capture under Gen. Zaru Bakshi[2] as related by Late Lieutenant General Moti Sagar is under:

"An important part of the Haji Pir complex was Bedori feature, it dominated the approaches. The main attack could only succeed if Bedori was captured first. This task was assigned to an all class battalion of the 19 Punjab. The NW approach lay along a very narrow ridge, not providing even a platoon frontage. The feature was sheer rock, no trees, no bushes, and not even a blade of grass. There was absolutely no cover. Artillery fire support was by one mountain gun which had been manhandled up with great difficulty—it could only fire over open sights.

The enemy bunkers had been tunnelled into the hillsides. The start line was at a height of 2750 meters, Bedori was at 3750—a 1,000 meters climb. The gun engaged the enemy. At last light, the Punjabis assaulted with two companies. The fighting continued all through the night, against a well-entrenched enemy. It was hand to hand fighting, but they kept on going with stubborn and indomitable spirit, from one bunker to the next, winkling out the enemy. They did not falter, in the 8-10 hours of darkness, never stopping, regardless of wounds and casualties, the climb was achieved, on the rocky slopes, full of loose stones, difficult of a foothold. With courage in their hearts and hope in their eyes, those brave and determined men captured Bedori by 0600 hrs. The job was completed by a column of the 1 Para Battalion, under Major Ranjit Dayal[3]. It involved a steep climb, over slippery slopes, they were exhausted, but with spirits soaring high, the Pass was taken by 1030 hours on August 28. Dayal earned a well deserved MVC.

The importance of this feature to Pakistan–has been highlighted by General Niazi in his book published in 1998. He was commandeered to take charge of the Sector of 12 Div at Muzzafarabad, the capital of POK (Pakistan occupied Kashmir). The Haji Pir Pass in Indian hands posed a grave threat to Pakistan, as it dominated the area, and posed a threat to Muzaffarabad.

The War- India having shield away at the Rann of Kutch – Ayub expected a break through, by a Blitzkrieg attack by the Pattons supported by fighters. India fought back under the leadership of Shastriji, with full participation by the people of the Punjab. When additional troops like Monty Palit's Division were moved in, they were guided by truckers knowing the way. 'Thela' drivers delivered ammunition to the gun positioned at the front. When we had to give up ground, they behaved most stoically. Army reciprocated; a gunner told me, "When we moved back to Asal Kot the fields were full of ripe chilies. " We harvested and filled them in gunny bags for our farmers. To prevent thefts I went to Taran Taran and bought locks and chains for their homes. "I personally remember the effusive hospitality offered on the trains. " Alas, it petered out south of Delhi. Our units were under strength, but in good heart. They were highly trained, as a result, a most surprising fact of this war, as recorded by Jogi[4], was that despite thousands of vehicles being driven on all types of roads and dirt tracks without lights at night, there were no major accidents.

On September 1, Pakistan opened with operation "Grand Slam" against Chhamb. Aimed at the capture of Akhnur, the gateway to Jammu, possession of which would stop access to Kashmir. A proverbial "chicken's neck" on which a lot of blood has been spilt. It had also been fought over in 1947. Due to UN restrictions, we had not built any fortifications, bunkers or, pillboxes. The lay of the land was such that Pakistan could exploit its superiority of armour and artillery. It, therefore, had little difficulty in slicing through our defences. Both, our Brigade and Divisional Commander were new to the area and the troops under them, making Pakistan's success a mathematical certainty. "Grand Slam Operation" opened with the main attack by a squadron of Pattons, supported by a motorised Infantry Battalion. Behind them were whole of 7 Division, two regiments of Armour and the whole of Corps Artillery, which brought down a heavy bombardment from 0345 to 0630 am. It was followed by sporadic firing till 0830. Our squadron of 20 Lancers in their Shermans reacted sharply and inflicted a number of tank casualties. General Choudhari, the Army Chief, asked the Defence Minister's Committee, for assistance, from the Air Force. The Air Force launched an air strike, within an hour, using four Vampires, another four followed. These old crocks had little chance against the Pakistani jets. Three of them were shot down, the effect was debatable. It was a most unprofessional way of waging war;-inter service co-operation has to be pre-planned. The thrust was held. The decision to hand over the operations to General Yahya Khan at this juncture came to our rescue. Although our counter attack on the 6/7th; without overwhelming artillery, armour and air support, failed. The Division. however successfully held Akhnur.

In the air Pakistan Airforce tried to rule the skies with their missile firing Sabers. We were lucky to have got into manufacture of the tiny fighters, Gnats, and had over a hundred Gnats in the air-flown by a bright lot of young pilots, like my son and his cronies Sukhi and Mukhi who had joined after the Chinese War. I had been associated with the production of fighters, the *"Saber Slayers"*[5] since their first appearance at Farnbrough Air Show in 1954. These single engine planes had a fantastic rate of climb, and small silhouettes that allowed them to approach, enemy planes unseen. They also had a very tight turning circle, thus could run rings round the Sabres. There were difficulties in production; snags were removed with the help of veteran pilots like Air Marshal Gole. Our pilots were great, they would provide an umbrella by flying high above the attacking planes. Then suddenly swoop down on the Pakistani Sabres. The name of the two Keeler brothers Trevor and Denzel is high amongst Sandhu, Pathania and Patney who were also decorated. The first Sabre was downed on September 2. The next followed two days later; after it developed into a routine. We also had Hunters, and three Squadrons of Canberras. MIG 21's had arrived but were not in service. A Mahavir Chakra was won by Jaggi Nath for a deep photographic recce. Both the Air Forces, without any special bombs, concentrated on interdiction–they attacked each others airfields—but to little effect. A few Vir Chakras were won for shooting up trains. Mukhi at Halwara told me of a strange phenomena. A flock of geese at the base took no notice of our pilots landing or taking off but made a shindy whenever an enemy plane approached the Airbase. Pakistan air dropped their Special Services troops, in groups of 50's to infiltrate our Air bases at Halwara, Adampur and Pathankot. They were to destroy aircraft and create panic. With no means of escape–they were soon rounded up.

INDIA threatens LAHORE—Appreciating the difficulty of restricting operations to Kashmir front, the Prime Minister had already authorised a general offensive. General Dhillon's 11 Corps crossed the International boundary at 0400 hrs on September 6. By 1000 hours, 15 Division was astride the Grand Trunk road, with its leading troops on the outer fringes of Lahore. They were led by their Commanding Officer Lieutenant Colonel Desmond Hayde. The 3 Jats of 46 Brigade crossed the border, carried on across the Ichhogil canal, into Pakistan territory, at Dograi. It was an epic effort, but no way could they push on to Lahore—without a pre-planned follow up. Pakistan, Air Force retaliated. The Jats had no other way but to pull back. The canal bridge was demolished by the Pakistanis. Baba Bhide[6], Chief Engineer 11 Corps records that during the brief interval, when there were no troops there, the buffalo population of Amritsar had increased by some 4,000!

Our troops had obsolete maps, which showed Bata chowk, as Dograi village. Across the canal, these showed open fields instead of huge complex of buildings of Bata Nagar with tall chimneys. On our way back from Lahore, I had passed through this congested area, with a War Memorial, in memory of the 16 Pathans and the Pakistan Rangers.

DOGRAI AGAIN—On the night of September20/21st -our gallant Jats, under the guidance of Brig Niranjan Singh commander 46 Brigade-supported by the Gurkhas and another battalion in Phase I—of the battle; with artillery fire support recaptured Dograi from the 16 Punjab (Pathan) of Pakistan. With great élan, they went in with their war cry "*Jai Bhagwan—Jat Balwan*" and fought with great bravery. Dograi was in their hands at 0330 hrs on the September 23. Besides their normal weapons, they used, bottles filled with Molotov Cocktail, which lobbed over the galli walls, exploded with a flash and a bang. The Jats won a number of bravery awards, including three Mahavir Chakras. It is to the credit of the battalion of just over 500 men, that they caused nearly 900 casualties. Took 108 prisoners, including the Commanding Officer and two officers of the 16 Punjab, and the battery commander. It has been glamourised in a book "Battle of Dograi" by Lieutenant Colonel (later Brigadier) Desmond Hayde[7]"

The Divisional Commander of 15 Division got involved in a bizarre episode, while swanning near the front; he lost his transport and diaries. Pakistan radio blared, their contents. There are conflicting reports of the incident. He was removed from Command.

Pakistan launched its main offensive on the night of September 9/10. Its 1 Armoured Division outflanked our 4 Division and captured Khemkaran. Their success was due to our not knowing siphons and viaducts, on the Ichhogil canal. The armoured division was in the forest of Changa-Manga near Kasur. It sallied forth and surprised the 13 Dogra battalion by suddenly emerging through a passage (Siphon) under it. They panicked! Being a Mountain Division, there were no RCL and anti-tank guns with the battalion at that time. Not understanding the enormity of the task on September 11 Major K.N. Singh's 100 Field Company was placed under 9 HORSE and tasked to destroy all six crossings over Ichhogil canal. When Pakistan troops had already crossed it over to our side in strength on night 10/11 September and even on 6/7 September Grenadiers were not able to secure Balanwala and The Punao Crossings which were strongly held from the other bank.

In an amusing reaction, Lieutenant General Katoch -the Colonel of the Dogras–came to 11 Corps, to inquire! A rumour spread, that Army HQ had sent him, to replace the Corps Commander! Part of "witch hunting" during war!

To regain Khemkaran, story goes that the 4 Sikhs, were egged on by their Colonel Commandant Lt. Gen Harbaksh Singh V.C to attack an unreconnoitered area to commemorate—Saraghari day, September 12. The Battalion was pulled out from Barki and assigned the capture of Khemkaran. They advanced in the dark, towards the objective, to marry up with their armour. At the break of dawn, when they thought they had arrived at their rendezvous—a voice on a loud hailer announced "Halt you Sikhs—you are surrounded by Pakistan Armour—lay down your arms—or you will be annihilated." There was no way the catastrophe could be prevented, a lone Major Valladaris, escaped.

Pakistan's plan was to capture the Grand Trunk (GT) Road Bridge over river Beas which would give them a straight run to Delhi. The force was directly controlled by GHQ and the 11 Infantry Division and 1 Armoured Division commanders were not on best of terms with each other. Their 1 Armoured Division was commanded by General Nasir Ahmed an Infantry officer not rated high–who could not take advantage of the Patton's night fighting capabilities, and was later removed from Command. The Division started at Asal Kot a small village near Khemkaran. In absence of information by foot recce from the Infantry due to excellent camouflage by the defenders and believing that their Pattons were indestructible; they joined battle as of yore. Leading tanks with flags and pennants fluttering raced across in broad daylight. They were on target to the Beas Bridge. This was confirmed by the "Op Order" recovered from Divisional Commander's tank which was disabled. He with his recce group drove right into the killing ground of 7 Infantry Brigade. They presented a target to Havildar Abdul Hamid of the 4 Grenadiers, who behind his Jeep mounted RCL gun, knocked out the leading tank. Then changing position sent another one up in flames. By then the enemy spotted and fired at him with machine guns–wounded him yet he shot a third tank.

Mortally wounded, he was awarded the highest award for bravery—a "PVC." Injured Nasir Ahmed was rescued by his escort; however, the Brigadier Artillery of Pakistan Division was killed. The advantage of longer range, was lost due to the chosen ground, with tall sugar-cane crop which restricted the fields of fire of the Patton tanks, with stereoscopic range finders and sophisticated computers. General Dhillon[8], Corps Commander a graduate of Thompson Engineering College, had no doubt used his skills in the selection of the battleground. Khemkaran distributory bridge was demolished by Major K. N. Singh of 41 Field Park Company under orders of Lt. Colonel Vaidya, commanding 9 HORSE and embankments the breached to flood both sides of the road as enemy armour was approaching. This gave one precious night to defenders. The flooding of the area started on the 9th. Thus restricting the advances by armour along Khemkaran-Patti road. Our defensive position, had river Sutlej on the east, an anti-tank ditch on the north-east, and was securely held by the two Brigades of 4 Division, with the extensive minefield and Nalas presenting tank obstacles covered, by the 2 Armoured Brigade.

Pakistan armour's frontal attack was a brash action, in an area limited by anti-tank obstacles. The tanks outstripped the lorry borne infantry, and were channeled into fields of tall sugarcane, where the Centurions of the 3 Cavalry in their hull down positions fired their highly effective 20 pounder AP (Armour Piercing) shot. Infantry used their newly issued R C L guns, by observing the waving sugarcane. Another factor to the detriment of the Pakistan armour was the inflammability of the Pattons. There is an Israeli riddle-"What is the difference between a Patton and a Ronson lighter?" To which the answer is, "Nothing except that the Patton lights first time. "Many tanks were abandoned due to the fear of the tank crew of getting burnt by fire which is 'Haram' in Islam.

General Dhillon told me of the visit of Shastriji. He took him to see four Pakistan tanks shot up in a two-acre field. He, then, lifted him up by the armpits on to a tank which he enjoyed immensely. By the time the President Radhakrishnan visited, 97 tanks had been collected in a field near Bikiwind, which became known as Patton Nagar. He recorded the inimitable obituary, *"Born in Detroit, died in Bikiwind."*

A controversial Pull Back advice had been confirmed by Captain Amrinder Singh Maharaja of Patiala, who was then ADC to General Harbaksh Singh. He had received the call from General Choudhary addressed to his GOC in C. Advice to pull back behind the River Beas was given in the early hours of the night of September 10/11. Harbaksh who had just returned after inspecting 4 Division rightly refused to entertain any such thought. In any case the ignominy would have been on his head—no one would have remembered the Army HQ suggestion! Western command had no reserves left, the Chief wanted to play safe as Munshi's 23 Division had not fetched up.

General Pat Dunn's Corps fought the biggest Tank Battle in Asia since the Second World War. Over 1,500 tanks took part in the battle, in the Sialkot sector. Our 1 Armoured Division led by General Rajinder Singh (Sparrow)[9], in his Centurion—"*Ankush,*" was pitted against Pakistan's 6 Armoured Division. On September 8, our 1 Armoured Division attacked from Jammu with three regiments, mounted on Centurions—the Poona Horse, (17H), Hodson (4H) and the 16 Cavalry.

The tanks outpaced their supporting infantry, and clashed with the Pakistani Armour, which had similarly outrun its infantry. The tank battle raged for 15 days. General "Sparrow" tasked Lieutenant Colonel Tarapore's 17 Horse (Poona Horse) to drive a wedge between Phillora and Chawinda. As a preliminary to the attack on Phillora, 400 tanks took part in the battle of Phillora. On September 11, while attacking Phillora, his tank was hit by an RCL shot; despite being wounded he went on and the formation carried the objective. The town was captured by the afternoon; an officer took the sign board "*Phillora Police Station*" to the General.

Colonel Tarapore was at the head of his regiment, during the Divisional battle for Chawinda on September 14 and further attacks on September 16, when he was mortally wounded. He was awarded a posthumous Param Vir Chakra for his outstanding valour. Poona Horse earned the Sobriquet "Fakar-i-Hind"-Pride of India from their adversaries. For three days, tank fought tank in a battle of attrition. The Pattons with all their gadgets were far too complicated for their Pakistani crew. Our tankmen were trained in the three round battle range technique. As soon as the Commander said "Sabot"—a solid armour piercing (AP) shell shall be loaded and the gunner would fire three rounds, one at the estimated range, next above it and the third below it—thus they usually registered a minimum of two hits. The final score claimed was 67 Pak tanks for loss of six! The smaller number of Centurions destroyed, can be attributed to their ability to take punishment; and enemy's lack of AP ammunition. As for Pakistan's losses, General Gul Hassan who later commanded their 6 Armoured Division, said that two of the Armoured regiments had to be issued new tanks. He also writes of the planned counter attack by the 6 Armoured Division, supported by Pakistan 1 Armed Division less 5 Armoured Brigade, which was moved by rail from Khemkaran sector to Sialkot and their arrival was awaited; therefore, General Gul Hassan managed to get the C in C to agree to a postponement of counter attack.

The end of the war came within a fortnight. We started receiving messages through the US embassy for a ceasefire. Colonel Surendra Nath, my nephew, commanding an Infantry battalion, noticed that the rate of enemy fire started falling after ten days.

Ayub's biographer records that their tanks ran out of AP ammunition in mid September, "Pakistan had imported High Explosive (HE) ammunition, instead of Armour Piercing (AP) ammunition. This mistake as Ayub mordantly said cost them the failure in this war".

On September 20, the United Nations demanded a ceasefire. Shastriji agreed despite opposition from the Defence Minister Y.B. Chavan. It was effective 0330 hours Indian standard time September 23.

P.V.R Rao, the Defence Secretary, told me of his visit to Jammu with L P Singh, the Home secretary. They went to look at the Sialkot battlefront. General Sparrow laid on a briefing for them at Chawinda. There was an elaborate display of maps, chinagraphs pencils and all—it looked like an artist's studio! The backdrop included a captured Patton trundling past for live effect! General Sparrow earned a bar to his MVC.

The Pentagon, fearing loss of orders for Pattons—initiated a study on its performance. It marked the end of gasoline-powered tanks.

This last war was between two Sandhurst trained Army Chiefs of Staff of their respective Armies—Musa and Choudhary. It will be remembered for being conducted as per rules of war.

At Dera Baba Nanak Bridge

Pak President Gen. Ayub Khan, the big blustering Pathan armed to the teeth by America decided to take Kashmir by the force of arms. He raised a second armoured Division, of Patton tanks, and got missile firing Sabre Jets. The stage had been set very carefully—India had shown tardy response to fighting in the Rann of Kutch where Pakistan Army had delivered a punch and accepted adjudication by a third party—the UK—there was no reason why it would not buckle under a BLITZ KRIEG was Ayub's belief.

In these conditions, XI Corps, defending the Punjab, under Jogi Dhillon had a difficult task. Here is an incident to show how exaggerated reporting by fighting troops can be neutralized by a good staff officer holding the hand of the Commander. In this case it was Brigadier Bhide, affectionately called Baba, the Corps Chief Engineer.

Dera Baba Nanak (DBN) is a salient in the river Ravi, where the boundary lies, half way across the bridge.

This is what happened at the bridge in Bhide' words:

"19 Infantry Brigade held the home side where 4/5 Gorkhas battalion was at the bridge site with two companies, in depth. The bridge had been prepared for demolition. Lieutenant Chandra Shekharan Nair of 63 Field Company had taken suitable position to execute his task should the situation so demanded.

On the night of September 5/6, this battalion came under heavy fire or so they say—heavy artillery bombardment. The forward companies of the Gorkhas pulled back, as the area had been accurately registered by Pakistan artillery. The Sapper firing party knew nothing about it and they stayed in position.

The next morning the Brigade Commander, who was directly under the Corps, rang up saying that as the Paks have crossed

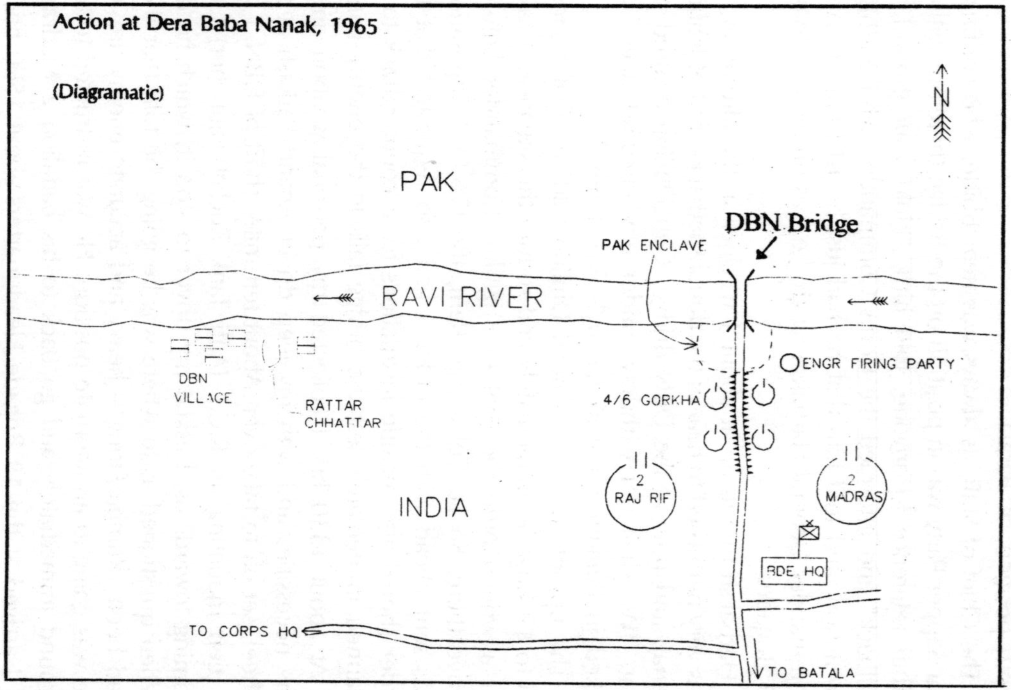

Action at Dera Baba Nanak, 1965

(Diagramatic)

over the bridge, occupied the position of the Gorkhas, and since their armour is expected across, should he now pull out to the depth position, at Batala."

The Chief of Staff, as always was near Bhide, who confirmed that Sapper Party was in position on the Bridge therefore without futher reference to anyone, told him, "Hold your ground, no pullback" and informed the Corps Commander, who endorsed the action, and told him that he had just heard from the Army Commander General Harbaksh Singh, he was on his way to visit 19 Brigade.

Harbaksh turned up at about 1 o' clock in the afternoon. He was very perturbed to hear of the double setback—first at Ichhogil Canal, and now at the DBN Bridge, Gen Dhillon assured him that, "We shall not withdraw under any circumstances." He asked in Punjabi "*Hun Ki Karan*?"—What next?

The upshot was that General Dhillon told me that I should go to 19 Brigade, immediately and restore the situation. I asked for a written order be sent to Brigadier Commander before I arrive there. So Lt. Col Chajju Ram, the G.S.O. I, (operations) was sent ahead with the order. I had also requested that this order should also include instructions for a counterattack to be planned immediately lest the bridge falls in the enemy hands.

At about 1330 hrs, I collected my personal weapon and a few necessities and with my jeep driver Anand Parkash at the wheel, set off to take over. About ten miles short of DBN, I saw a jeep mounting an R.C.L (Anti Tank Rocket Launcher) Gun, coming towards us. I asked my driver to stop it, which he did. When questioned as to where was he going, he told that there had been "Rumba Firing"—heavy and accurate enemy fire and he was going to an alternate position. He was instructed to turn around immediately and go back to his battalion 2 Madras.

I arrived at the 19 Brigade Headquarters about 1500 hrs, to find that the atmosphere was fairly cool. The Brigade Commander had got his second wind and a counterattack had been planned by the depth battalion, the 2 Rajputana Rifles. They were badly under strength, as the rifle companies had hardly, 50 to 60 men. Therefore, a two-company attack was planned from the

West to the East—along the riverbank to capture the foothold established by the enemy on our side of the bridge.

The attack went on fairly late in the night and was successful. The enemy was about a platoon strength and taken by surprise. In fact when the assaulting Company Commander who was a subaltern, on reaching the objective found the enemy Company Commander with the signal operator and runner standing around. The enemy Commander shouted, "Who's there, shoot them." Sure enough our officer was quick to draw and shot the Pak officer and the other two. Now the battle was over as the platoon fled across the bridge.

Throughout the night confusing and alarming reports were being received. At 2 o'clock, the 2 Madras on the right reported that a Squadron of Pattons was about to attack them. Knowing that there were no Pattons in the area, they were asked, "How was it that enemy armour posed a threat at their position?" They reported that the tanks were not far away as they could hear their engines. They were ordered to stick to their position firmly.

The Brigade had under its command a Squadron of Shermans, under Major Ravi Inder Singh of 14 Horse. Now these tanks were blazing away their guns at something or the other, we could not make out at what. Every 10 to 15 minutes, he would turn up in a highly excited state and report—we have got one more Patton shot. This went on for some time, until the tally was closer to 15 enemy tanks destroyed. Our armour had no night firing capability, therefore all this sounded vague.

Throughout the night there was firing of small arms and explosions of artillery shelling and typical "plops" of the mortars. I happened to be the only one who was capable of judging the type of fire based on the volume and sound having been under fire in World War II.

Next morning at about 4 o'clock, it was being considered whether or not to demolish the bridge–when a mighty bang was heard from the other side, the Paki's had blown down the bridge from their side.

I went up the observation tower and found that only one big span of about 200 feet had been demolished and as the river here was pretty shallow—if required we could still get across. A patrol under 2 Lt. Nair was sent to check the correct position. Nair went across, and finding an unclaimed enemy jeep brought it along. He was awarded a Vir Chakra for his various acts of gallantry earlier as subsequent to this foray.

The reconnaissance of the previous nights of the battleground revealed that there were undoubtedly movements,—by some Chaffee enemy tanks in the area. However, they had been driven away by the sheer volume of firing of our Scinde Horse tanks. Two Chaffees were found abandoned and a third lying at the bottom of an embankment, with its tracks up in the air. Obviously its driver had driven it off the embankment—where it had been left.

None of the three tanks had a direct hit on them from Ravi Inder's Shermans. It goes to show that the assessment of situation in night operation must be taken from the incoming situation reports which may be inflated.

By about 11 o'clock, I was convinced that there was little further action left for me to retrieve the imaginary bad operational situation and therefore got back.

General Dhillon asked me about the state of the Brigade Commander. On my reassurance that he had totally recovered. The Corps Commander turned around and remarked;

"Yes, you always tell the truth don't you ?."

A Windfall

Second Lieutenant M.A. Lakshmanan of 36 Field Company of 9 Engineer Regiment (Madras Sappers) was sent on a task of demolishing a stone wall which was holding up our advance to Sialkot. It was suspected that some Pakistanis weapons were dug in, in its vicinity. Lakshmanan had gone out on a night patrol, with the Gurkhas to locate the offending wall. It was not all that formidable "A stone wall in mud mortar, along the side of a field". He made a quick mental calculation, doubled the charge, and returned with Naib Subedar Shanmugum and two sappers, along with the explosives.

It was an early fresh and slightly misty winter morning. Just as the first light shown on the wall, his party laid the charge, got back to his pre-selected cover point and pressed the exploder. There was a loud explosion with a lot of dust, a deafening bang, stones flying all over. Hardly had the dust settled down, when he was still assessing the damage and taking advantage of the freshness of the morning breeze with deep breath. Suddenly, he noticed some movement in the tall grass on his left. Looking closely he realised that—it was from a sandbagged parapet of a trench.

He was neither equipped and trained nor in an adequate position to tackle the dug in enemy. The safest course for him was to have beaten a hasty retreat and reported completion of the task, but he was made of sterner stuff. Intent on keeping up the Sappers traditions, he decided to act. He took the logical action and got his party to take up prone positions and ordered them to fire at the offending trench.

Absence of return fire lifted his courage. Soon he was surprised to see a lone figure approaching him, with a white cloth, which he was waving, about to make it obvious that he was intending to surrender. The huge explosion and the fire from the semi automatic rifles had unnerved the enemy. Lakshmanan with his three rifles and with only a Sten gun in his hand was apprehensive—was it a Ploy?

What to Do? There were no set procedures—he could think of nothing else except to shout *Halt!*

Which he did and the Apparition did halt in his tracks immediately.

The next command, "Advance to be recognised" (In his best Hindustani) followed logically. A huge middle aged VCO with a lurid red moustache in Pakistani uniform came forward saying "I and my party of 18 other ranks of the 10 Baluch Regiment—surrender."

This was a real bonanza. Lakshmanan despite the small size of his force—literally as well in actually—took heart and ordered the Baluchis to come out of their trenches without their arms and fall in behind their leader with their hands up. The enemy party soon complied. He ordered his squad loud and clear to keep them covered by fire and shoot to kill in case of any infringement to his orders.

This very tense drama had been going on for quite some time—but our friend Lakshmanan, who had been overly busy, figuring out-what to do next–had no time to inform his Company Commander.

At last he passed the message on his intercom, "Have captured nineteen POWs." His boss was incredulous, "How the Hell?" He shouted, "REPEAT your message."

Lakhshmanan repeated it loud and clear. Overjoyed at the news, he said, "Well done-hold on-reinforcements on way."

While waiting for the reinforcements—the time was passing too slowly—they heard the noise of tanks clanking away in the distance.

The Baluchi's also heard it—their faces lit up optimistically, they murmured, "*Hamare Hain*" (These are ours), there was a tenser atmosphere as they grew restive but they did not dare to jump the three automatic rifles menacingly pointing in their direction.

Soon the tanks were with them—not their but ours. The tankmen were surprised to see the bag of their Sappers. It was a troop of 7 Cavalry Reconnaissance Regiment, leading the attack on Shakargargh. They took charge of the POWs.

Now that the offending wall was no longer there they went through like a knife in butter in no time. The story went the rounds of the Division. It was on every one's lips. Our 2 Lt. Lakshmanan became a celebrity overnight!

To check the authenticity of this story, I wrote to the Commandant Madras Sappers and Miners—and received a copy of the unit's "War diary" which records that this action took place on the Sept 9, 1965, at Octroi post on Jammu-Sialkot road (Map Ref 842279). Brigadier M. L. Khetarpal who was the Chief Engineer during the 1965 war remembers the incident.

Notes

1. Lt General M. M. Khanna M. V. C.
2. Lt General Zorawar Chand Bakshi PVSM, MVC, VrC, VSM.
3. Lt. Gen. R. S. Dayal M. V. C.
4. Lt Gen J. S. Dhillon, Padma Bhushan, GOC XI Corps
5. Lon O Nordeen in his book *Air Warfare in the Missile Age*
6. Maj Gen V. B. Bhide, AVSM
7. Brig Desmond Hayde M. V. C.
8. Lt Gen. J. S. Dhillon, Padma Bhushan
9. Maj Gen. Rajinder Singh Sparrow, MVC and Bar

11

Blue Helmets

This is about our mission to Congo. The huge country, located on the equator, encircled by the 4380 kilometer long river Congo, which drains the forests of central Africa into the Atlantic. Till, recently it was called Zaire, but now it has reverted to its earlier name of Congo. Its capital Kinshasa (Leopoldville) is at the mouth of the Zaire, the language of the elite and the government agencies was French.

Forty years ago, a large body of Indian soldiers was seen in Congo. What were they doing there? It was rumoured that they were there to conquer Congo to settle its population!

Wrong, it was a Brigade of our finest troops, in Blue Helmets/ *pugrees,* on peacekeeping mission: pitched into battle, at the request of the United Nations to prevent the disintegration of the country. They got involved in real fighting, left behind some 39 killed. An officer, won our highest bravery award, a *Param Vir Chakra.*

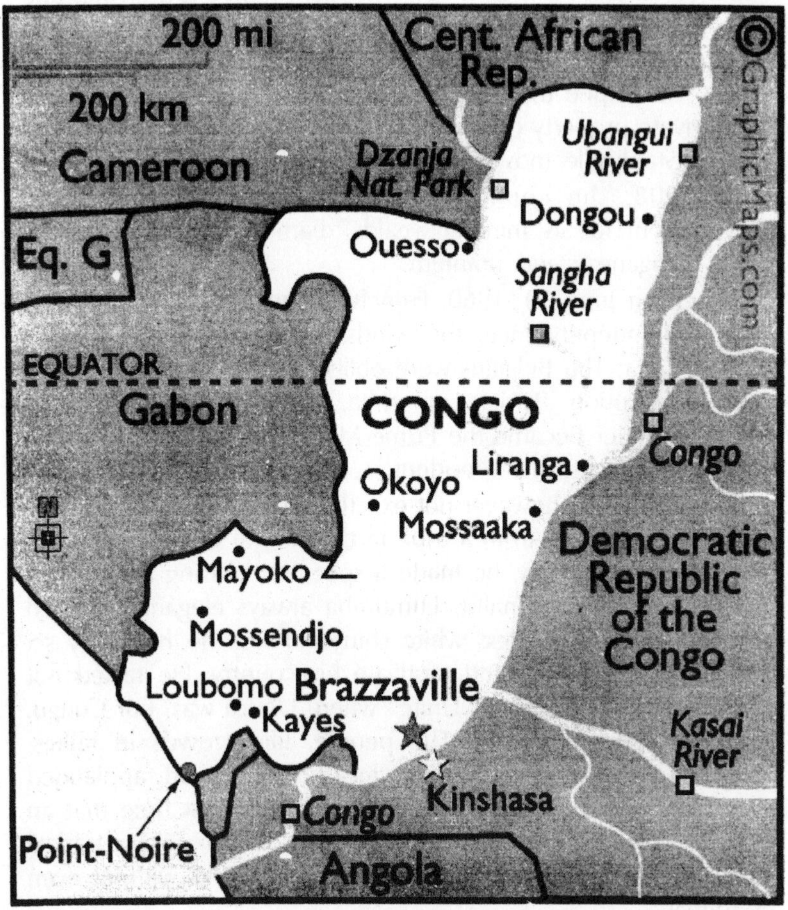
200 mi
200 km
Cent. African Rep.
©GraphicMaps.com
Cameroon
Dzanja Nat. Park
Ubangui River
Dongou
Eq. G
Ouesso
Sangha River
EQUATOR
Gabon
CONGO
Congo
Liranga
Okoyo
Mossaaka
Democratic Republic of the Congo
Mayoko
Mossendjo
Loubomo
Brazzaville
Kayes
Kasai River
Kinshasa
Congo
Point-Noire
Angola

The gigantic country is inhabited by small clusters of people living in hamlets scattered across the jungles -ten people to a square mile. Most of us have heard of the famous British explorer—Stanley—who in 1871 was sent to find Dr. Livingstone in central Africa, and his famous greeting, "Dr. Livingstone I presume?" Later, he travelled the length of the river Congo. Finally, he helped the Belgians to develop the Congo Free State as a private property of King Leopold II. A town on the river was named Stanlyville (now called Kishangani).

In 1908, the country was taken over by the Belgian Government for its mineral wealth, diamonds, copper, cobalt, zinc, manganese and uranium.

When on June 30, 1960, French Congo across the river was promised independence, the winds of freedom were blowing across Africa. The Belgians were obliged to follow suit. Under a new Constitution, Patrice Lumumba, (34 ex postal officer), the majority leader became the Prime Minister and Kasavubu (42 a clerk) was elected the President.

Lumumba was however not exactly the man they wanted, he was a fiery speaker from a tribe in the central Congo. Educated at Moscow University, he made a speech criticising the Belgian rule. A staunch nationalist, Lumumba always elegantly dressed in a suit with a spotless white shirt and tie—he had only six months to go round, and motivate his country. He talked not of tribes, but of Congo "l' Unite" where Uhuru was, our Congo, Victory, "independence." The people, like crowds in rallies, gathered as for a harvest festival, in a holiday mood, applauded him with fervour in their eyes. It was for the first time that an African voice was raised. For hundreds of years slave traders had shipped millions of slaves all over the world without even a squeak.

When the Belgians granted independence, they expected that it would possibly have a few ministers, with large bungalows and big cars and that the business would go on as before. The business was in any case, in the hands of a company called Union Ninie're du Haut Katanga—it mined cobalt, copper in the mineral rich southern Katanga. It provided more than half of the revenues, and called the tune. Elizabethville, the capital city, with roads lined with jacarandas, was run by the company.

The sudden grant of the independence, left Congo like a ship, hit by storm and abandoned by its crew. The administration was in a hopeless state as there were hardly any Congolese officers; Colonel Mobuto (29-ex army clerk) was the senior-most soldier.

At the capital Leopoldville, the Congolese Army mutinied with heavy-handed imperialistic intrigue. The Belgians flew in their paratroopers, ostensibly to protect the less than 2 % European population. They speedily brought the situation under control. Well-armed mercenaries joined the Katangese Gendermerie, a force that was better than the Congolese Army.

Within eleven days Congo was threatened by cessation of the prosperous mineral rich region Katanga (Shaiba) under Moise Tshombe, a grocer by trade but from a rich tribal family.

The President appealed to the United Nations for help. The U.N. deprecated the secession of Katanga, and by its resolution of July 13, called on the Government of Belgium to withdraw its troops from the Congo.

Stanleyville province on the river and Albertville on Lake Tanganyika also broke away, but later reeled back. The main problem was Katanga, which obdurately stuck to cessation. Central Government employees were deported from the province on July 18. A new green, white and red flag of independence was hoisted.

The matter was raised at the U.N. The non-aligned nations supported Lumumba. Russians on an appeal from him threatened to send their troops or volunteers. With Kennedy coming into the White House, the West also changed. Unity of the Congo became the U.S. policy, and reconciliation was in the air.

The United Nations voted to establish a U.N. peace keeping force. Rajeshwar Dayal (Indian Ambassador to Pakistan) was appointed to head the U.N. operations in the Congo.

The U.N. troops were introduced personally by the Secretary General Hammarskjold, who, despite of Tshombe's objections, flew in with four plane loads of Swedish troops to Elizabethville on August 12. A battalion each of the Irish and the Swedes followed. At first India sent medical cover but later when it was clear that Katanga was set on breaking away and started to recruit mercenaries, and openly expand its armed forces, Pandit Nehru agreed to send the Indian Army. The maintenance of the unity of the Congo was one of the UN's primary tasks.

The Belgians, thwarted in maintaining control, promoted their protégé Colonel Mobutu, and tried to eliminate Lumumba, who was protected by the UN troops, but all roads leading to his residence were blocked. He decided to escape and fled to the East towards his own tribal people. On the way, however, he was apprehended by the anti-national troops and brought back to the garrison town of Matadi. Here, his charisma spread beyond the prison walls, the local people began to agitate on his behalf. Thereupon he was flown to Katanga, where he was murdered.

The problem festered for two and a half years. Congo became the largest peace keeping operation by the United Nations. It involved 19,825 soldiers from 30 countries including Argentina and Yugoslavia. India provided the largest military contingent. Police forces were lent by Ghana and Nigeria. Three Indians—Narsimhan as Chief of Mission to the UN Secretary General, Rajeshwar Dayal and General Diwan Prem Chand—played an important role.

In view of the urgency, Indian Army's 99 Brigade Headquarters, 3/1 Gorkhas, 1 Dogras and 2 Jats were airlifted in USAF Globemasters. 13 Field Company Madras Sappers, traveled across the Indian Ocean aboard a US Naval ship, on April 1.

Heavy equipment and vehicles followed. The Brigade group, some 6,500 in number, and a detachment of the Indian Air Force, a troop of Armoured cars, and a battery of four inch mortars, were under the command of Brigadier K.A.S. Raja.

The battalions of the Swedes and the Irish were very hard pressed and outnumbered by the Katangese Gendarmerie, the main danger being at Katanga. The Gorkhas and the Dogras were sent there. The Jats were stationed at Leopoldville, the capital, to watch over the volatile Arme Nationale Congolese

(A.N.C). Raja started by securing the airport and Elizabethville. His forward HQ was at Kaming airbase.

I am indebted to Mr. M.A. Rehman of the Indian Embassy for this part of this story. He had joined the IFS after having served in the 8 /12 Frontier Force Regiment in Burma during the war. One morning, he had flown to Kaming about 35 miles from Elizabethville, the capital of Katanga. He was a most welcome guest at the Brigade mess. Brig Raja's mess was about 9 miles from the town. From there two roads branched off, one to the town and the other to the camp of the Irish Battalion.

The road junction was held by a small detachment of the Dogras. All of a sudden the mercenaries attacked them to disrupt the communications. They used a monstrous contraption, a bulldozer, protected by Armour plate welded around it. The young officer incharge of the post fired a rocket from the RCL (Rocket Launcher)—the dozer's armour disintegrated. It was left naked. The mercenary troops on either side disappeared into the jungle.

A sharp barrage of 3-inch mortar shells fired by the Dogras made sure that they did not show up again. The road to the Brigade tactical Headquarters on the outskirts of the city was restored.

General Mekeon, the Force Commander, landed at the airbase. The two R's (Rahman and Raja) met him and were returning to the town in their staff car when they were fired at. The Brigade HQ was located in a substantial building in the town. It was subjected to intermittent fire from the insurgents. It did not cause much damage.

The matter was discussed with Brig Raja, who though circumcised by his orders as a peacekeeping force, could exercise his discretion for self-protection. It was decided to launch "Operation- Rumpunch" in August to expel the mercenaries. They were a motley crowd, armed with the semi-automatic F N rifles whilst our Jawans had old wartime bolt operated rifles.

The opponents also had a few Mercedes armoured cars. Major Padda, described as a hawk nosed-cavalryman, commanded our armoured cars. He took over Albertville, an important town and communication centre on the shore of the lake Tanganyika, and distinguished himself by capturing several gendarmes' positions.

During September 1961, the two companies of the Gorkhas under Major Thapa, by threat and manoeuvre, obtained the surrender of the Katangese in central Katanga.

Eventually on November 24, the UN declared that secessionist activity was unacceptable and authorised the Secretary General to use requisite measure of force.

When in early December, the Katangese attempted to establish a roadblock, Colonel 'Chotta' Hazari of the Dogras sent Captain Salaria of the 3/1 Gorkha, with a platoon to remove the threat. Salaria not only held the position alone for many hours till he was mortally wounded, his platoon stuck to its post till it was relieved. For his exceptional gallantry, Captain G.S. Salaria, who was killed in action on December 5, was given an immediate award of a *Param Vir Chakra* for his leadership and courageous defiance of the enemy for hours on end. He was its sixth recipient.

The only reverse the UN forces suffered was at Jodatville where a detachment of the Irish had earlier been surrounded by an exulting Katangese force led by the mercenaries and assisted by a jet trainer aircraft, a Fouga Magister. Their mastery of the air had prevented relief columns from using the roads. The UN was thin on the ground to go after the Katangese and chase them away.

Raja asked for more troops to be sent to Katanga as well as aircraft to combat the Fouga. Both calls were answered promptly. The UN General MeKeon had personally experienced the situation at the Brigade Head quarters which daily received about an hour's attention from the long range Katangese mortars. Their gendarme was being trained and their number was increasing. A Battalion of Ethiopians, a Canberra bomber under Squadron leader Soares as well as Swedish Saabs arrived in a matter of days. They were stationed at Kamina base.

A major operation *"Morthor* (Smash)" was planned on December 13. The Brigade took over control of the communications. It was reported that their major concentration was around a stadium in the middle of the town. A barrage of gunfire was laid down, the attacking troops brought up to the start line. The enemy decamped presumably during the barrage and few were found. This was a relief for the U.N. troops, who in any case wanted to avoid bloodshed. It established the superiority of our forces.

The operation was, however, only a partial success as Tshombe escaped to Nadola to his supporters in Northern Rhodesia. Nothing very much happened to change the overall picture of a separate Katanga. Hammarskjold on a visit to Katanga to arrange a cease-fire, was killed in an aircrash on his way to Nadola.

Norhona's Brigade

On the completion of its year's tour of duty, Raja's Brigade was relieved by an Independent Brigade group, under the command of Brigadier Norhona.[1] Major Sunderji[2] was his Brigade Major. Joglekar's 22 Field Company, part of the Brigade Group disembarked at Dar-e-Salam and arrived at Elizabethville at the end of October 1962.

They found the town almost under a siege. To prevent the mercenaries, having a free run, it was decided to establish a roadblock on Avenue Katuba to control their entry and exit. The task was allotted to" D" Company of the 4th Battalion of Rajputana Rifles.

The story of the "Road Block"as told to me by its officer in command, Captain Surender Nath[3], my nephew, is recorded as under in his words:

"A few days after the road block had been in position, one morning a press reporter arrived and asked if all was well. He further told me that the hotel porter had tipped him to cover the place. While we were still talking, a low wailing sound floated over to us from the Katangese side of the town. This kept on increasing, till we saw a crowd of women and children moving towards our roadblock."

"The crowd was chanting some sort of litany. The demonstrators encircled us; their rhythmic chant became louder. They started dancing and making obscene gestures, showing their bare bottoms to the troops at the barricade. Soon the noise rose to a crescendo—women became frantic—and started demanding that the brave Indian soldiers should give them -"*Mutotos*" (children). The situation was getting most embarrassing. The gestures were not only very clear—the women left nothing to imagination. It was also getting a bit dangerous—a hundred Jawans surrounded by about 10,000 women who were fully aroused by now. Suddenly, at this stage sound of firing was heard from the Katangese side of the border. Soon after the women started to withdrew. The press reporters who had gathered to see the "*Tamasha*," also moved away. As they were leaving I managed to have a word with Mr. Justin of the *London Times*, who along with the others had been tipped to proceed to the local hospital, for a news conference. My Company Commander asked me to check if by any chance any of the men had fired in self-defence. I did so and reported that all weapons were clean and that the ammunition was correct. By noon rumours were circulating that the Indian troops had fired on a peaceful group of women demonstrators and brutally killed them by bayoneting a number of them. We were shocked at the baseless allegations, especially as the troops had displayed discipline and restraint of a very high order in the face of extreme provocation. The Indian Commander asked for an independent inquiry by an international team. Three Judges, from the World Court of Justice at the Hague, flew in the same evening and by the next evening I heard the broadcast by the BBC. The team of Judges from the International Court of Justice, after a detailed 'on the spot' investigation, emphatically denied the involvement of Indian troops in the alleged killing of unharmed women and children demonstrators. The three dead bodies produced by the Katangese had been shot with 7.62 mm calibre weapons; the Indian Army used .303-inch ammunition. The bayonet wounds were found to be five inches deep and half inch wide.The bayonets used by the Indian troops were eleven inch long and one inch wide."

Whereas the firing incident was reported about 9 a.m, the autopsy report showed the likely time of death to be at least eight or nine hours earlier. Apparently bodies of three women patients who had died earlier at the hospital, were shot up, bayoneted and produced as the brutal handiwork of the Indian soldiers.

Major K. Sunderji, the Brigade Major, recounts several humorous stories of the expedition. I repeat an incident, which is a bit uncharitable to our friends of another nation, but was published in the "Hindu." Apparently the Battalion was ordered to capture a railway overbridge astride which the Katangese were entrenched. The Indian Brigade Staff was receiving messages of the progress of the attack on their wireless set. Suddenly, it received a message that the attack is being abandoned as the Battalion had run out of ammunition. On being asked as to where they were when the attack was called off, the answer was that they were still at the start point.

To the incredulous Indian Major it appeared strange that they had not moved forward. So he asked for a clarification. Their explicit reply was, "Yes, we did not move, we were attacking with Fire."

By December, it was clear that it was no use carrying on negotiations, it was decided to take action to defeat the Katangese, if necessary with force as sanctioned in the U.N. resolution of November. An operation code named "Grand Slam", planned and executed under the directions of Reggie Norohna and his Brigade Major Sunderji, was launched on December 28, 1962. The objective was to capture the rich mining towns of Kipushi, Jadotville and Kolwezi. The first action occurred between the Rajputana Rifles and the Gendarmerie who had placed a road block at the outskirts of the town. This was removed by a company of the Rajputana Rifles on December 30.

The Katangese had large stocks of explosives, with which they had prepared extensive demolitions on the highways coming into the capital. Captain Surinder Nath was the Mortar Fire control officer with the right hand leading company. Those days we had cumbersome manpack 22 sets with only five feet of cable from the heavy lead acid battery. The telephones were little better than the children's match box telephones. Despite the handicaps, we were soon on our way forward.

The Irish Battalion, under Colonel Delaney, advanced towards Kipushi, a copper mining centre near the N Rhodesian border. It was held up by a blown bridge over a deep stream. The engineers were asked to build a bridge, but there were no stores available. An officer had seen a Bailey bridge near the golf course. Recourse was taken to the only alternative-to steal it temporarily. Accordingly with great difficulty it was man-handled back over greased sleepers and launched within two hours of it reaching the site. The Irish captured Kipushi, by 1230 hours the same day.

Advance to the other centers on the axis Elizabathville Kolwesi started on the New Year's Day. It was led personally by Reggie Norohna who often rode ahead with only his swagger stick to receive the surrender by firm persuasion and bluff without bloodshed. For capture of Jadotville, Lieut. Hassan Muzzaffar of Bombay Sappers led the party—a Squadron of the 7th cavalry, and a column of the famous 4th Madras. It included Major Sami Khan and Capt. Ragu Shastri carrying a captured automatic rifle, much preferred to the bolt operated Indian rifle.

Suddenly, the advance was held up. Both the road and rail bridges over the swollen Lufira river with a 7-8 knot current, had been demolished. Ferrying operations to take the few jeeps across were not possible, as there was no support equipment available, not even a steel wire rope. Major Sunderji took off in a helicopter to see if there were any crossing places elsewhere in the area. Luckily, he found a ford where there were vehicle tracks, going across to an installation of the Union Miniere, the Belgian mining company.

Very boldly, he went in to check the depth and rushed back still wet. Reggie, meanwhile, was already waiting with a column of three jeeps -one with an Anti Tank Gun, another with a Machine Gun plus crew, and with his own personal escort. They were the first to get across the river Lufira in their jeeps.

Meanwhile, despite heavy rain, the Madras Regiment with the help of the Sappers had been able to get across on an improvised, foot bridge of drums and cordage.

When Reggie and party reached the Union Miniere installation, they were met by a delegation of the company officials and the Mayor of the town in a black suit and a gold chain of office who explained that he was in no way connected with Moise Tshombe. He offered the keys of town which were accepted. Thus, ended the affair of the Swagger Stick capture of Jototville on January 3. Bailey bridges were flown in on the 18th, thereafter there were no hold ups.

Finally, the town of Kolwesi was taken on January 21. Katanga was cleared of the mercenaries in a shortwhile. Colonel Z.C. Bakshi's 2/5th Gurkhas, with their '*Khukris*' were particularly dreaded by the mercenaries. For his leadership, in these operations, Zoro Bakshi won a VSM.

Tshombe was forced to eat the humble pie, and started to negotiate with Leopoldville.

Task Accomplished

Little known is a fact told to me by Brigadier Khanduri, that task was completed not without a melodrama. Whilst the Security Council had indeed approved the offensive, there were pressures a plenty on U Thant to go slow. It is said that a garbled message, "Operations Se finis,"was passed on to Diwan Prem Chand at the force HQ.

Unlike personnel of our contingent, many were happily enjoying generous UN pay and alowances. So much so, that in one of the battalions everyone received the pay of one higher rank, consequently there were no privates, only Lance Corporals.

While the Brigade was employed on UN operations, the Chinese attacked Indian defences in NEFA. The task having been accomplished the Brigade to a man wished to return, to fight for their country. The arrangements for transportation back to India of our troops took a long time to finalise, by then the Chinese had retired from the Indian territory in NEFA. When at last the

Brigade was all packed up to depart, they received a message that the luggage allowance per person would be reduced to 100 kg. from 120 kg.

Movement tables had been prepared with meticulous staff work: each person's kit had been weighed. The redrawing of the loading tables and a counter-order at the last minute was not acceptable to the Commander.

Major N.L. Berry,[4] the Brigade interpreter in French who also helped the Force Commander, was detailed by the Brigade Commander to proceed to the Logistic H.Q. of the U.N at Leopoldville and resolve the issue. Arrangements had been worked out on the basis of the move of the Brigade as a whole, without having to leave a rear party.

Major Berry(Tanny), who had worked hard with the Brigade staff in the planning, flew there and was told by the Canadian head of the Logistics, and Major Jamil, his Pakistani deputy, that, there was "No Choice". With the backing of his Commander, Tanny was equally adamant, and said that in that case the Indians would stay on.

Matters came to a head, till in the afternoon, Logistic HQ relented, it was agreed to airlift the Brigade to Mombassa, from there it was to go back to India by ship.

That evening in deference to our past relations, the Pakistan contingent at Leopoldville, arranged a mess dinner for Tanny. Next morning, he flew back, mission accomplished.

Congo was not all fighting in big game hunting. Our head doctor was a passionate hunter. He went and shot a wild elephant in one of the national parks. There was talk of his having to be sent back in disgrace.

Rahman, our Charge de Affairs, went to see Foreign Minister Bomboko and remonstrated, "After having risked and often lost their lives for the Congo, surely a mistake can be forgiven."

Bomboko said it was very difficult but he would do his best, "You see the problem is serious as he shot the animal under the statue of King Albert."

Tailpiece—The Brigade traveled from Mombassa by *U.S.N Blathford*. Whereas half the ship's hold was allotted to the Brigade, the other half was taken up by the baggage of the platoon or so, with its five officers, the providers of logistic support. The unkindest cut was that the officers had all decided to drive back in their new Mercedes.

Mobuto was President for decades till the year 1997, when he was forced to abdicate, and he died after looting several billion dollars. Zaire is again called Congo.

Notes

1. Lt General Reggie Norohna M.C and Bar
2. General K. Sunderji
3. Maj General Surinder Nath A.V.S.M, S.M
4. Maj General N.L. Berry

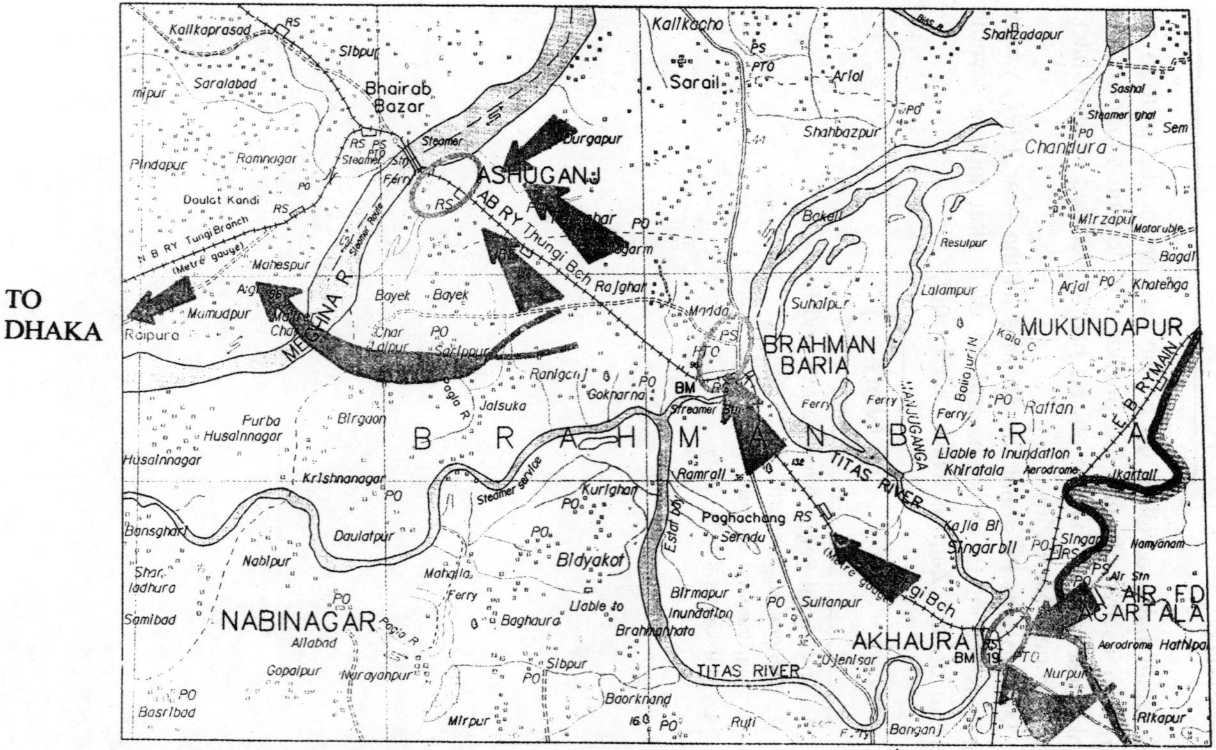

Map of East Pakistan—Central Eastern Sector
(Prepared from captured maps)

Race to Dhaka

(Boundaries not to scale)

❮12❯

Bangladesh War 1971

General Yahya Khan had taken over as the President of Pakistan in 1969. Sheikh Mujibur Rahman of the Awami League of East Pakistan, had won an overwhelming majority in the National Assembly. Mujib had promised that Bangladesh would be a Federal State, responsible for all matters except defence and external affairs. This situation was acceptable neither to the military ruler nor to Bhutto of the Pakistan Peoples Party (PPP) in the West. Decision to impose Urdu as the national language, further aggravated the situation. A stalemate ensued. Yahya Khan visited Dhaka and decided that the best way would be to suppress the Bengalis with force. On March 25, the postponed date for convening the National Assembly, he instructed Lieutenant General Tikka Khan, the Martial Law Administrator, to launch "Operation Searchlight," and left for Rawalpindi.

Tikka Khan arrested Mujib, and started to sort out the Bengalis. Sultan M. Khan, in his book *Memories & Reflections of a Pakistani Diplomat* published by the London Centre for Pakistan Studies, confirms the massive repression. He quotes Bhutto (in an interview with Oriana Fallaci), "During the night of March 25, 1971, only 50,000 and not a 100,000 people were killed in Dhaka." It was a massacre, a "*Qatal-ae-Am.*" Even in World War I, notorious for its high casualty rate, there were never ever 50,000 casualties in a day. To say that these occurred during pacifying the unarmed civilian nationals, is beyond comprehension. Khan goes on to say, "I heard Generals talk of starving East Pakistan into submission, cleansing it of the non-Muslim population, or injecting some good Muslim blood into East Pakistan as a solution to the problem." Further he adds, "Yahya himself lived in cloud—cuckooland—especially after sunset when neither he nor his cohorts were ever sober."

No wonder that they misread the message from China as conveyed by Dr. Kissinger, that "China will not be an idle spectator." Sultan and Gul Hassan, the CGS, went to China on April 9, to seek Chinese support. Sultan records, Zhou En-lai advised that it was inappropriate to discuss East Pakistan affairs in the presence of Ambassador M.M. Kaiser, a Bengali. Later, when Bhutto wanted him to be put on board a PIA plane, and be sent back, they declined.

The massacre in Bengal started an exodus of population to India. Eight million refugees, mainly Hindus, crossed over by August. India appealed to the world community, to have this aggression vacated as the refugees imposed a massive economic burden. Two additional Pakistan Infantry Divisions were transferred by sea and air to East Pakistan. General Niazi, the overall Commander, then disarmed the Bengali military units, the paramilitary forces and the police. This led to the formation of an insurgency organisation—*Mukti Bahni,* under Colonel M.A. G. Osmany, (later the chief of staff of Bangladesh). India took full advantage of the unprecedented unrest and supported *Mukti Bahni,* to foment guerrilla activities.

The involvement of the armed forces of India started with Sam Manekshaw being asked to attend a meeting of the Cabinet in April 1971. The political leaders were very concerned with the influx of refugees and wanted him to take offensive action. He was told to prepare for military operations quickly. In his brutally frank manner he told them about the poor state of equipment—particularly the tanks, and advised that it would be suicidal to operate in that part of the country during the monsoons. After the meeting, Indira Gandhi detained him to discuss his problems. Manekshaw refused to be stampeded, he obtained authority for single point provisioning under General R.G. Williams, the Director of Inspection, and bought adequate time to move men and materials. It was agreed that the earliest hostilities could commence was after the monsoons. According to her biographer Pranab Roy, Indira Gandhi gave him an excellent political briefing of limiting hostilities within a timeframe of three weeks.

While the troops and materials were being moved, there were many a border incidents in which the *Muhkti Bahni* took an active part.

India's first victory has neither been officially chronicled, discussed or assimilated.

Masterly handling by the Chiefs of Staff

Despite the antiquated control of the Ministry of Defence and dismal background of 1965, the service chiefs, Sam Manekshaw, S.M. Nanda and P.C. Lal met daily, to discuss operations.

The Chiefs had direct access to the Prime Minister as mentioned in an incident related to me by Admiral Nanda, and mentioned in second part of this chapter.

War started when the Pakistan Airforce staged an "*A La Israel*" attack on our Western airfields, at dusk on December 3. Luckily, P.C. Lal had anticipated their timing without night bombing aids –it had to be on a moonlit night. Our planes safe under hard tops, suffered little damage.

Air Marshal Dewan of Eastern Air Command set up a tactical command HQ alongside General Arora at Calcutta, and fully supported the Army. Air Commodore Zaheer, Air Officer Commanding Agra Base, took charge of interdiction, bombing and transport tasks.

A major air force interdiction was at Karachi. IAF's Hunters and Canberra caused havoc at their main port of entry at Karachi. It cost us a Canberra.

Pakistan had borrowed ten F-104 from Jordan of which four were lost.

The climax of war was an attack by IAF on the Governor's residence at Dhaka on the December 14, it hurried the end of hostilities.

For the first time the Navy took part. In a daring operation, it sent missile carrying Motor Torpedo boats for a raid on Karachi harbour on December 4, which caused considerable damage and started fires in the oil installations. India fired USSR, S N–2 Styx missiles against Pakistani targets.

"Ghazi", a Pakistan Navy submarine on loan to it from the United States, caused us greatest concern. An intercepted message from Karachi to it –"*Vikrant* was at Vizag"—indicated that it was lurking around waiting for the aircraft carrier; however, Ghazi was sunk on December 3. The sonar of one of our destroyers, coming out of Vizag, heard a hard echo and it asked Delhi for instructions. It was told to fire its full pattern of depth charges and went on its way for its task to Chittagong. The first victory of the war thus went to the Navy—when fishermen picked up a life jacket with the U.S. markings.

The aircraft carrier *Vikrant* lived to launch 160 sorties against naval targets and airfields at Chittagong and Cox's Bazaar.

A "*Mujib-Bahni*" was activated by General Uban, a Gunner like Wingate, was a born leader of men. He came from the wild Hazara district of the Frontier and had been brought up with fire arms—as a loader of his father's rifles to repel attacks by dacoits.

His first step was to travel widely and make friends with real leaders of Bangladesh. He enlisted four student leaders, Tufail Ahmed, Abdul Razak, Sheraz Alam and Fazalul Haq Momi—a nephew of the Bangla leader Sheikh Mujibur Rahman. Operating under the Ministry of Home Affairs, he raised a 7,000 strong *Mujib Bahni* which operated in conjunction with the special Frontier Force, which had been raised to deal with unconventional war situations. This Force 22 was commanded by Uban.

The first to take off were Uban's *Mujib Bahni* from Demagiri in Assam. They entered East Pakistan's Chittagong Hill district forests, well before the formal war. Indira Gandhi had told him,

"You may act as you like but I shall deny my Government's involvement if you are caught". Uban sat on the hills overlooking a camp of Pakistan Rangers on the far bank of the Karnafuli River. After a study of their routine, he sent a platoon strength of his stalwarts across the river in 6-8 men assault boats to get behind the enemy post, and mount a surprise attack. The main force crossed over when the enemy was already on the run. They feasted on the food left by the Pakistanis. All at the cost of only one casualty. The problem of getting medical aid was solved by his pilot Jamis Ji, who on being contacted by wireless, asked that the location be indicated by lighting a fire with a lot of smoke. By lunch time every house had a fire. Jamis took a risk, landed, tied him in a bed sheet and saved him.

The PM on hearing of it chided Sam for being slow. The army had only been making, probing attacks.

Hostilities started on December 3, when Pakistan Air Force attacked our western airfields. Three Corps of eight divisions surrounding East Pakistan then commenced a squeezing operation. The IAF gained air superiority within four days.

57 Division Crosses River Meghna To Dhaka

In the East, 4 Corps sector Major General B.F. Gonsalves was the Commander of 57 Mountain Division which operated from Agartala. He was opposed by Pakistan's 27 Inf. Brigade commanded by Brig. Saidullah, which had the 12th FFR and the 33 Baluchs.

Operations started early in the morning of December 3, when 14 Guards attacked Gangasagar Railway station. The Pakistanis offered stiff resistance, the Guards suffered casualties. Havildar Albert Ekka of the Guards won the first and the only PVC in Bangladesh for conspicuous acts of bravery at the cost of his life.

One of the most imaginative and daring exploits was that of Maj General B.F. Gonsalves, Commander 57 Mountain Division.

After the capture of Akhura, the rail and road centre opposite Agartala, by 8 a.m. on December 5, he found that Pakistan had lifted one of the rail tracks and were using it as a road. Gonsalves thereupon got the Corps Commander General Sagat Singh to change his plan. On December 9, our 73 Bde pushed on to Ashuganj. Gonsalves wanted to capture the bridge over the Meghna intact—but this was not to be. The bridge was blown by the enemy. The Pakistani's were pushed across the river and were bottled up in Bhairab Bazar from where there was no road to Dhaka.

Thereupon General Sagat Singh arranged the Air Force help to lift 4 Guards by MI 4 helicopters and land them on the west bank. Major Samsher Mehta (Shammi) with his P.T.76 tanks swam across the stately Meghna river.

Thus surrounded on December 16, operations closed on Pakistan. On the riverine eastern sector, with wide rivers, it was PT 76 Russian tanks which won the day, by swimming across the three km. wide Meghna River – towards Dacca, the capital. It reminds me of a historical event in 1956, Marshall Zukov and General Thimayya visited the CME to witness a river crossing. We were embarrassed as one of our pontoons sunk. Zukov remarked, "*Our tanks swim*"—little did we know that 15 years on, our tanks of 63 cavalry would also swim.

For the first time, we used heliborne forces, directed by Major General B.F. Gonsalves of 57 Mountain Division, a para trooper and a pilot, who showed great flexibility. Advancing from Agartala, in the east, on the capture of Akhura, the transport junction, he found the enemy withdrawing.

There was an eight storey high rice silo at Ashugang, which provided an observation point to Brig Saidullah waiting on the east bank of the Meghna with a number of troops waiting to go across—he personally led a counterattack against 18 Rajputs and caused some 120 casualties.

However, enemy panicked and ordered the demolition of the only bridge on river Meghna. By December 9, our troops took upon defensive position at Bhairab Bazaar to isolate the enemy on the west bank–with no road link to Dacca. Gonsalves decided to leave the enemy in this town watched by his troops.

General Sagat Singh met and discussed with Gonsalves at Bhairab Bazar (BB) on December 9. He used his dozen MI-4 to fly across his troops, from the stadium at BB to Narsingudi-Raipura in 110 helicopter sorties. Another innovation was that the Airforce was asked to carry 1000-pound bombs on the conveyer belly of each of the two Caribous and drop them on Bhairab Bazar at first and last light each night. It kept them out of the battle.

The final action was on December 11, when Major Mehta swam his tanks across the Meghna. The Pakistani's gave up on December 9. 4 Guards and a troop of 82 Light Regiment were airlifted at 3.30 p.m. in 10 helicopters to Raipura area across the Meghna, giving a clear way to Dacca. Captain Sirohi of the Air OP led the fight. 4 Guards moved to Narsingudi the last water obstacle on December 10. From the stadium at Brahman Buria, General Gonsalves arranged for 110 sorties to be flown in 36 hours. Flying continued till the moon set and again from first light at night helipads were marked by torches without their reflectors. Meanwhile, Grover's 56 Field Company performed the difficult task of repairing five steamers and sent these from Anantpur to Gokarnaghat. These steamers moved reinforcements across the Meghna opening the way to Dhaka.

In the north, Lieutenant General M.L. Thapan, seasoned Commander of 33 Corps, told me that as usual our Intelligence was poor. Pakistan had fielded the equivalent of four Divisions against us; the fourth had been deployed into East Pakistan by air and sea via Colombo without guns and heavy equipment. There was a heavily defended and well-dug in Pakistan Brigade at Hilli. An attempt to dislodge them cost a battalion—75 killed and double the number wounded. He, therefore, ordered that the defended positions be bypassed. The move was so successful that after the war the Pakistani's complained. It was not cricket!

Surprisingly, Dhaka, the capital of East Pakistan, was not designated as the principal target. Troops of 95 Mountain Brigade from the north (101Comm zone), under Brigadier Kler, a Signals Officer, were one of the first to reach Dhaka. Kler had picked up Pakistani internal messages from Islamabad to General Niazi at Dhaka to surrender at midnight December 15/16.

The fighting ended about 4 p.m. on December 16. The surrender ceremony was held at the race course, in the fading light of the day. Those present, included Air Marshal Dewan, Vice Admiral Krishnan, General Arora, his wife, Generals Sagat Singh and Jacob. The photographs had to be amended to keep the lady out of the picture.

The biggest MFU (Military Fuck Up) was an unplanned Combined Operation. The confusion was great; Naval wireless jammed Army wireless. For the amphibious landing on the coast at Cox's Bazaar to cut off Pakistani forces, a hastily gathered force sailed, without the Naval contingent which failed to arrive. The place selected was unsuitable for beaching the LSTs which had to stay about a 100 metres offshore, troops were asked to wade across and two drowned eventually, *Mukti Bahni* ferried them across in fishing boats. Two days later, when they reached Chittagong in local lorries, the war was over. Incidentally, the road to Arakan was blocked by *Mujib Bahni* who destroyed it taking advantage of a bomb hit on the Karnafuli river bridge.

The Birth of Bangladesh: General Uban told me some snippets about the establishment of its new Government. He recalls that when he landed at Dhaka, there was such a large crowd to receive him, that he was almost carried on the shoulders of his men to the Hotel International. Mr. D P Dhar and Mr. Kaw, were already there wanting to have a new Government.

Uban went to meet Mrs Mujib at Dhan Mandi. She was emphatic that there should be no Government till her husband returned. Uban, with only one thought in his mind of his Guru—Sita Ramdas Baba Onkar Nath of Bengal—flew with his friends to Calcutta. Beseeched his Guru, true to his word, he said, *"Durga Mai Mujib ko Chor de-uske bagair kam nahi hoga."*

Soon after Mujib was released by Pakistan and was flown in by India to form a Government. But few know that he asked for Uban, a Sikh with a turban and all, to stay as his advisor, for almost a year and a half. Till Uban retired.

Bungled Peace

India had taken some 93,000, prisoners of war. With little interest in military matters, I doubt, if anyone in our Defence Ministry set up had ever read the Duke of Wellington's advice on the treatment of prisoners. In 1804, he had said:

"When war is concluded, I am definitely of the opinion that all animosity should be forgotten, and all prisoners should be released."

Ignorant of matters military, we hung on to the POWs, fed and paid them for almost two and a half years. Pakistani prisoners were quite different to the highly disciplined Japanese. There were problems galore—Commodore Khanna told me how while transporting the POWs from Chittagong to Calcutta he found them indisciplined. A Pakistani Lieutenant Colonel was down right abusive and refused to salute Indian officers.

Tindi[1], who was involved in housing them in the Central Command, told me, two escape stories. The first was a proper tunnelling job at Fatehgarh, the home of the Rajput Regiment in northern U.P. There were about 200 officers in a separate enclosure, they made a crawling tunnel which was lit by electric lamp to prevent claustrophobia. One evening there was an unusual excitement in the camp when people were asking the "Escapees" to take messages for their loved ones. The security officer made a special visit but could not detect anything out of the ordinary. He left telling the *sentries* to be more vigilant. That night three escaped through the tunnel which led into a deserted shop some 30 yards outside. They got home via Nepal but the Sapper officer who designed the route was left behind.

Major Ibrahim, an old soldier, was held in Agra Central Jail—he had been there also in 1965. One evening while playing bridge, he said that he was going to the loo. His friends got worried when he didn't return, they started looking for him.

Soon the whole camp was looking for friend "Burro." He had discovered a tree close to the fence which he had climbed and got out through the drain under the wall. Thereafter got lost in Agra bazaars.

There was no point in our keeping, feeding and looking after them. It would have been prudent to return them to their homes immediately after the cessation of hostilities, reducing the feelings of animosity as, after all, we have to eventually live side by side. No advantage was taken for their long retention and subsequent release during the Shimla summit.

The excuse for delay, by the theoretical decision-makers, was that they wanted to consult the Government of Bangladesh on punishing the guilty. Eventually, nothing happened on the war crimes. Bangladesh was more interested in being recognised as a state.

Lt. General Bhagat, VC, the Army Commander Central Command, personally ensured that the prisoners received proper treatment. We sent them back, I hope, without much rancour. All were provided new clothes. General Niazi left on April 30, 1974.

We had returned the captured territory in the Sialkot and other sectors. It was said that we did this to prevent the displaced people from suffering.

This time Pakistan claimed that the territories captured by either party on the LOC (Line of Actual Control) should be retained. It thus retained Chammb, and we were left in possession of a few isolated villages in the Shoyk river valley in Ladakh.

"Chammb" the scene of fierce fighting in the three Indo-Pak wars where scores of our gallant men laid down their lives so that we may live today, is no longer our land—it is now called Iftakharabad. We won the war but presented it to Mr Bhutto, as a peace offering.

There were no Generals to advise the Indian PM. General Candeth, the Northern Army Commander, told me that the nearest he ever got to the seat of power was an occasional drink with D.P. Dhar.

Mr Bhutto, the Pakistan PM, and his delegation was due to return to Islamabad without a settlement next morning. When suddenly on July 2, 1972, he asked for a personal meeting with Indira Gandhi. This meeting fixed for 6 p.m. was held without any aides. At the end of this one-hour long meeting, Bhutto announced that an accord had been reached. Their main point on which there were still problems was the sanctity of the LAC (Line of Actual Control). Bhutto was later tried and hung—his daughter Benazir who had been with him is the only survivor who can throw any light on whether there was any secret agreement. So far as the Army is concerned Pakistan will in future not have to attack the town of Chammb. A phrase "*Without prejudice to the recognised position of either side*" introduced as a peace offering dogs Indo-Pakistan relations!

Western Sector—in 1971

Chammb was attacked again and our planners were surprised that the enemy came back to attack the same place. It was an obvious target to cut off Kashmir from the rest of India.

This time to drive a wedge through to Jammu, a massive assault was launched by a Division supported with a Brigade of Armour (150) tanks and guns (180), more than were used in the whole of Bangladesh.

The attack started with an artillery bombardment at 2030 hrs, according to Major General Jaswant Singh, the enemy went through the entire depth of our position within 70 minutes. The fighting was the fiercest, he had ever seen since the Second World War. The enemy overran two or three of our forward screen positions, but next morning, four of our posts were still held despite the severe assaults.

The next three days were spent on attacks and counter-attacks; till we decided to make a readjustment of our position and move to a better defensive position behind the river Tawi. Pakistani's tried to cross the Tawi on the 9th and 10th and secured a lodgement on our bank of the river. We replied with a strong counter-attack, and evicted them by the night of the

10/11th. The water of the Tawi turned red with the blood of the wounded, as some of the Pakistani troops who had been issued cards saying that they would go straight to heaven if they were killed in battle kept on coming on in waves on their way to "Jannat".

Havildar Khazan Singh of the Guards went across the Chammb Bridge over Tawi, with a pack load of six missiles and took up a position. He saw two tanks at a distance of about 700 meters, shot both of them, as well as another sheltering in a defile, and blew up it's rear portion. He was awarded a Vir Chakra, for his excellent marksmanship. Lieutenant Colonel P.K. Khan won a MVC as his Sikhs killed over 500 enemy. Major Pannu won a posthumous Vir Chakra, Captain Kaushik, and others won recognition.

The Chicken's Neck, south east of Chammb, is a jagged piece of Pakistani territory, the area between the Akhnur River and the Jammu Tawi. It is shaped like a dagger pointing towards Jammu. The defence of this particular piece of territory could not possibly be in better hands than that of my neighbour, Major General Zorawar Chand Bakshi, the most decorated officer in the Army. Zorawar had trained his division in leadership, by personal example—his motto was "Follow Me." The story is told of an officer who was asked to capture an enemy sepoy for identification. The prisoner, who was injured in the struggle after he had been questioned, made an unusual request that he be allowed to meet his captor. When confronted by an officer he was flabbergasted—in his army officers did not go out with patrols, certainly not with fixed bayonets.

This area is much larger than the Chammb area; our attack went in on the December 5, the full moonlit night with fair visibility. Launched from the rear of the enemy positions, they were taken by surprise. They could neither get reinforcements nor find escape routes. The enemy ran in front of the ferocious assault. Lieutenant Colonel Rawat, commanding a Gurkha Rifles Battalion, recounts how they kept together in the attack by carrying 7,000 yards of tape, they advanced 8,000 yards into the enemy territory without a single man losing contact with his sub unit. They were amidst thick *sarkandas*, 15 feet tall, when they came under MMG fire. One of the Naiks[2] went in and charged the enemy—who ran, as they had no guts. By 1600 hrs as the tanks arrived, he mounted the company on the tanks and they kept on advancing till we were out on the far side of the Neck.

One of their smart Captains used two *charpoys* (string beds) from the local village, as stepping-stones, one at a time for leap frog through the minefields. The Pakistan Air Force in close support used MIG-19 which carried out three runs but on the fourth, he was shot down with LMG fire by Riflemen Dhan Bahadur Rai, who won a Vir Chakra. Our pilots worked extremely hard and achieved a record of six sorties each day. Some of these were in support of the ground operations. Flight Lieutenant Sahota and Syad Iqbal Ali won Vir Chakras for shooting up tanks in the Chammb sector. The Keeler brothers and others were decorated for shooting down the Sabres. The highest award, a PVC, was won by Flying Officer Nirmal Jit Singh Sekhon who undauntedly attacked a formation of six Sabres over Srinagar and downed two before he was shot down.

Admiral Nanda told me how during the second week of the war, he happened to walk into the PM Indira Gandhi's room, at a moment of history. As he opened the door he found the lady deep in thought in front of a map. Not wanting to interrupt her he started to withdraw, but was invited to come in. On inquiry, he found that the deep thought concentration was on Army's request for permission to attack Lahore. Indira's thinking was somewhat on the following lines—*whatever we gain by conquest would have to be returned. Whatever else, there would be heavy casualties on both sides. She as a mother would have to answer—was it really necessary to make the number of widows?*

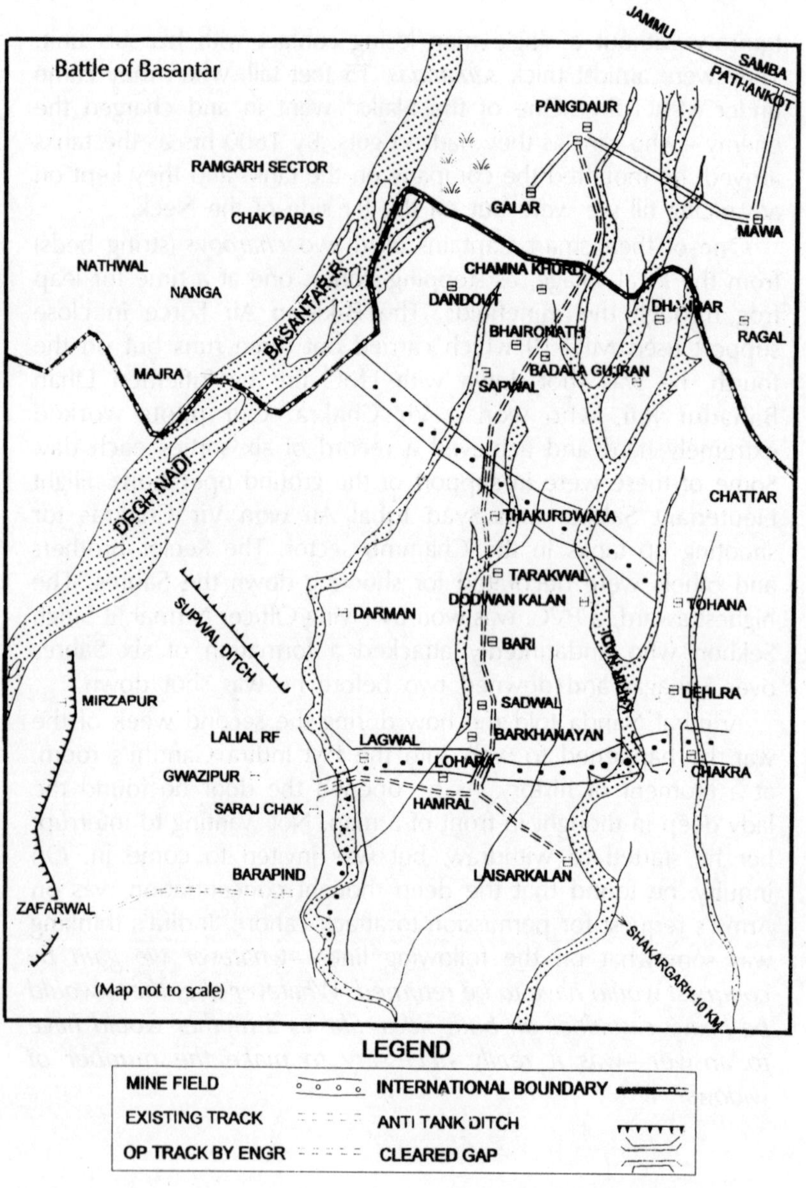
Battle of Basantar
JAMMU
SAMBA
PATHANKOT
PANGDAUR
RAMGARH SECTOR
GALAR
CHAK PARAS
MAWA
NATHWAL
CHAMNA KHURD
NANGA
DANDOUT
DHANDAR
BASANTAR R.
BHAIRONATH
RAGAL
MAJRA
BADALA GUJRAN
SAPWAL
DEGH NADI
CHATTAR
THAKURDWARA
TARAKWAL
DODWAN KALAN
TOHANA
SUPWAL DITCH
DARMAN
BARI
DEHLRA
MIRZAPUR
SADWAL
LALIAL RF
LAGWAL
BARKHANAYAN
GWAZIPUR
LOHARA
CHAKRA
SARAJ CHAK
HAMRAL
BARAPIND
LAISARKALAN
ZAFARWAL
SHAKARGARH-10 KM.
(Map not to scale)
LEGEND
MINE FIELD
INTERNATIONAL BOUNDARY
EXISTING TRACK
ANTI TANK DITCH
OP TRACK BY ENGR
CLEARED GAP

The Basantar River Battle—December 15-16

See sketch- Plan of minefields to be cleared prior to Tank battle with Pakistan's 8 Armoured Division. Two P.V.C's were won.

This tank battle saw Second Lieut Arun Kheterpal on a Centurian and Major Hoshiar Singh of the Grenadiers winning the Param Vir Chakras, our highest bravery award.

I am thankful for the description of this part of the operations to Colonel (Later Lieutenant General) B.T. Pandit who commanded the 9 Engineer Regiment and evolved the mine breaching techniques for deep minefields put in by the Pakistanis. Those days the Trawl tanks as well as bridge laying tanks (BLT)—three of each type were, on the War Equipment Table (WET) of the Armoured Brigade. When 16 Armoured Brigade led by Brigadier Vaidya joined 54 Division commanded by General Pinto, they agreed to place the trawls and the bridge laying tanks (BLT's) under command of the Sappers. From then onwards, a special team was built up for the breaching of the deep minefield, under Major Chowdhary commanding 405 Field Company. The first actual task took place on the late afternoon of December 6. Chowdhry's team wanted to test the procedure by day light breaching, but there was a troop of enemy Chaffee tanks on the far side of the minefield. The Squadron Commander of the Hodson's Horse had the answer—he took half of his squadron about two miles away and opened fire on the Chaffees. Pakistan armour was thus engaged and allowed Chowdhary to put his Trawl tanks into the minefield and enabled him to carry out his trials. After this Trawls managed to get through, he put the second tank with an overlap and followed it up with two tanks. The tactical concept of selective breaching and assault lanes was born.

The Battle of Basantar took place on the night of December 15/16. The Sappers, under Major Vijay Ratan Chowdhary, started trawling a safe lane through the deep minefield. They were held up at times. Naib Subedar Doriaswamy and Captain Ravinder

Nath Gupta, by their daring reconnaissance, facilitated the advance. Both were awarded Vir Chakras. Gupta was honoured posthumously. The whole regiment of Poona Horse on their part sent their "C" Squadron through the minefield, in pitch darkness while the lane which had not yet been fully marked. This in the face of the fact that the Centurions had no night fighting devices; it was an act of great courage and determination.

Colonel B.T. Pandit then commanding 9 Engineer Regiment told me :

"The sight of the tail lights of the 13 Centurian tanks at Basantar river going across the minefield gap shall remain the best sight in my memory."

They overran a position occupied by the Baluchis of Pakistan on the far edge of the reserve forest. Most of the Baluchis had fled at night and a few that remained to offer resistance were eventually made prisoners. The history of the Poona Horse mentions an enemy JCO, whose hands were tied by a rope and secured to the engine deck of 'Kooshab' the commandant's tank, who on being given a cup of tea said, *"Aaj ki larai se yah sabit hogya hai ki bari takat se kabhi bhi takkar nahin laini chahie."* (Today's action proves that it is wrong to confront a great power.) Later under the cover of smoke screen, the enemy was repulsed with tank casualties on both sides, clouds of smoke were seen going up from the gasoline fuelled Pattons on fire.

While this engagement was going on, a threat developed in the Jarpal position. Captain Malhotra (Mallu) of "A" Squadron, of Poona Horse—with 3 Troop under Capt. Avatar Ahlwat and 4 Troop under the baby of the regiment, Arun Kheterpal, was asked to help. They charged headlong towards the enemy with their guns blazing. They came under fire by RCL guns, they charged and overran them, and at pistol point captured their crews. Pakistani tanks under Major Naser made a determined attack—he attacked with his whole squadron of Pattons. Avatar's tank was hit, and he was seriously wounded. His driver pulled back the tank and managed to take him to the RAP (medical post). Soon after Mallu's tank gun developed a mechanical defect, he asked for permission to withdraw, but the CO issued instruction to hold fast. Arun was thus left as the sole contender in the field. His tank also received a hit, luckily it ricocheted off. His tank received a second hit and caught fire. Mallu ordered Arun to abandon his tank, but Arun citing the hold fast order by the Commandant, and the critical situation, replied, "My gun is still functioning, and I will get these bastards"

He systematically began knocking the enemy tanks, ferocity of fighting was intense—four enemy tanks were hit, the last tank to receive a direct hit was that of Naser, the Pakistan commander, barely 75 meters away. Arun's tank withstood a fourth hit, this time it went through the turret porthole, killed Sowar Nand Singh the operator, mortally wounded Arun, and severely wounded the gunner Nathu Singh. At this stage Sowar Paryag Singh showed great courage, he evacuated Nathu Singh to another tank, which had come alongside. In this process, he too was wounded in his leg, but undeterred, he withdrew the tank under cover, put out the fire, and started to brew tea to give it to Arun. Soon, Naib Risaldar's tank came alongside and they tried to extricate him but he collapsed and died. Arun was killed fighting in his tank and won the Param Vir Chakra. Major Naser also received Pakistan's "Sitara-ae-Jurat." Interestingly, it was Major Naser who looked after Brigadier Madan Khetarpal, Arun's father, during his recent visit to Pakistan. He was full of praise for the guts shown by his adversary. *Century of Warfare* by Messenger and published by Harper Collins, gives the losses as Pakistan 45 M-47 Pattons Vs 15 Indian Centurions.

Major V.R. Chowdhary of 9 Engineer Regiment who had mastered the minefields, received a posthumous Maha Vir Chakra.

The Shakargarh battle is unique,- two Param Vir Chakras were won on the same day. Major Hoshiar Singh of the Grenadiers covered himself with glory in the Basantar river operations. His battalion was a part of 47 Infantry Brigade. It successfully established a bridgehead across the river on the night of December 15. He won a P.V.C for holding against the fierce counter attack on December 17 morning. The attack was led by the Pakistani Battalion commander personally. The fierce fighting resulted in the Pak CO and two of his officers being killed, our hero though wounded, lived to wear his well earned decoration.

Cease-fire orders followed at 1800 hrs the same evening.

Notes

1. Lt. General S.N. Sharma—later E-in-C
2. Corporal

❮13❯

Sri Lanka

This Pearl of the East or "Emerald Island" in the Indian Ocean was once an ideal holiday resort. Colombo, its capital, with its wide avenues and graceful tall palms gently swaying in the sea breeze, was mesmerising. It was turned into an inferno. This is the story of a medieval world of Buddhist monks, temples with bells and elephants, and how it changed.

In 1948, Ceylon, a Crown Colony, attained "Commonwealth Status." Bandarnaike, its first Prime Minister, owed much of his electoral success to the slogan "Sri Lanka for Sinhala only." The pressure of population, and internal strife, led to a demand to get rid of the Tamils. This created a complicated problem; they had been there for a long time.

The Tamil kingdom of Jaffna had existed unto the 16th century. It was first captured by the Portuguese, and then by the Dutch. Tamils of the east coast and some north central parts had been under the Sinhalese Kingdom of Kandy till 1815, with whom they intermarried. The British in mid 1880s brought in a large number of labourers for the tea plantations in the central highlands, they were called Indian Tamils.

In 1956, Sinhala was declared as the official language, and the discriminatory policies against the Tamils resulted in a sense of alienation amongst them. Bandarnaike was assassinated in 1959, his wife took over. She was in power when I passed through Colombo. She followed the socialist pattern—nationalized the industries, a number of Indians left the island. The economy faltered, I could not even buy a pair of canvas shoes at Colombo. Pressure on the Tamils increased, attempts of these people to make Sri Lanka their home was thwarted.

Sixty million people in Tamil Nadu have strong ethnic ties with the Tamils in Sri Lanka. In view of this, successive Indian Prime Ministers made efforts to alleviate their suffering. Talks between Prime Ministers Senanayke and Nehru led to the 1948 Citizenship Act. It allowed two years time to apply for citizenship. Innumerable difficulties were created, registration did not take place till 1962. Prime Minister Shastri entered into a further agreement in 1964, which recognised the problem of stateless persons. In 1973, Mrs. Indira Gandhi visited Colombo and signed yet another agreement with Mrs. Bandarnaike in 1974, by which Sri Lanka agreed to grant citizenship to 75,000 Tamils, and India agreed to repatriate an equal number. By 1983, some 50 million Tamils of Sri Lanka were declared as non-residents.

The United Nationalist Party, under Mr. Jayewardene, was in power when institutional political violence broke out. Sri Lanka went into flames; riots broke out on July 26, 1983. Tamil militant opposition had grown up under various names: LTTE, TULF, TELO and others, of which the LTTE (Liberation Tigers of Tamil Elam) was the dominant group. They attacked the police and military personnel. This time the Sinhalese outdid all others, in the scale and the brutality of their attacks. Homes and shops were set alight; service personnel killed political prisoners in maximum-security prisons. The retribution by the Tamils was not long in coming; how far these groups were supported by the Tamils of Tamil Nadu is a question which is unlikely to be resolved.

Since 1983, for four years it was clear that some sort of military intervention in Sri Lanka might be necessary but no planning had been undertaken. A few months before involving the Indian Military in an attempt to end the internal strife, and bring the country to some semblance of order and finally impose a possible solution to the ethnic war was a lesson on *"How not to do It."* A highly secret contingency plan was made at Army HQ, with participation by Navy and the Air Force. Southern Command was told in June that Gen. Depinder Singh, its G.O.C in C, would be the Force Commander. Luckily, the Corps of Signals was able to set up a communication plan. The situation at Jaffna, the home of the Tamil revolts, had been difficult for a long time. Ceylon government cut off all supplies to this area. The genocide and the economic blockade made India come to the help of the beleaguered Tamils.

An unarmed flotilla carrying supplies was challenged by Sri Lankan Navy. The first intervention by India was an airdrop of relief supplies. Five A.N 32 planes flew in on June 4, 1987. Photographs in Mr. J.N. Dixit's, our High Commissioner in Sri Lanka, book *Assignment Colombo*, published by Konarak, show supplies being distributed.

There were discussions at Delhi. Prabhakaran's, the head of the LTTE, visit to New Delhi was not all that happy; it is said that he was confined to his hotel room.

The induction of an Indian Peace Keeping Force IPKF, was decided on July 27 at Colombo. Jayewardene the 82-year old President of Sri Lanka, was a well meaning person. Even before the signing of the agreement, he asked for early positioning of Indian troops at Jaffna and Trincomalee. He also asked for help to airlift his troops out from Jaffna. Accordingly troops of the Army Reserve Division, under General Harkirat Singh, from Secunderabad were flown in.

Rajiv lent two A.N 32 planes to lift the Sri Lankan troops from Jaffna which delayed his return to India by two days. Whilst inspecting the guard of honour at his departure at 10.15 a.m. on July 30, a Sinhalese Naval rating attacked Rajiv Gandhi with the butt of his rifle. He escaped with a bruise.

It showed up the travesty of the "Accord". The Prime Minister of Sri Lanka, Premadasa, and the Buddhist monks opposed it.

The whole exercise was unplanned. No steps had been taken in indoctrination of the IPKF troops. There was not even a primary pamphlet for the troops on the land and people of Sri

Lanka what to say of maps which should have been issued before sending in the army. We hadn't learnt from our mistakes as pointed out by General Jacob, during Bangladesh war. Some 500 maps given to the IAF for delivery to HQ, IPKF were later discovered lying undelivered in a hangar at Palam.

There was not even an embarkation unit to monitor the dispatch of troops. The Commandant Madras Sappers Bangalore told me, how a skeleton embarkation staff was hastily put together after the first mix up. The disembarkation was pathetic. A contingent arrived at Trincomalee in two vessels, one an LST, the other a Scindia ship.

I recall our arrival on Admiral Soman's ship *Jumuna* in 1952, when Trinco's beautiful harbour, was a second home to the Indian Navy. Now our ships had to ask a private firm, for permission to use its only jetty. The ships were received by the LTTE.

No thought had been given to command and control of the IPKF. It was placed under Madras Area, which in turn reported to Southern Command at Pune, and finally to the COAS at Delhi.

Co-ordination with the Navy and the Coast Guard for transportation of troops, and the control of the shallow waters in the Palk Straits, was through Naval HQ Vizagapatnam.

For getting a helicopter from the Air Force, to evacuate wounded soldiers, reference had to be made to A.O.C-in-C Southern Air Command.

Lt. General Sardeshpande was appointed to command the IPKF, as he had added a footnote to a routine letter of thanks to the Chief, "I wonder if there is a place for me —in Sri Lanka or Ladakh". Within a week, he was on his way to Jaffna. It was a limit of adhochism.

Further, it is interesting to note that when the Prime Minister visited Sri Lanka to sign the accord, no military staff accompanied him.

If one is to believe a responsible civil official, General Sunderji is supposed to have advised that the Indian Army can neutralize the LTTE in a fortnight or at the most in three to four weeks.

The Indo-Sri Lanka accord was signed between Prime Minister Rajiv Gandhi and President Jayewardene at Colombo on 27 July, 1987. The stated aim was "desiring to preserve the unity of Sri Lanka, recognizing the areas of historical habitation of Sri Lankan Tamil speaking peoples and preserving it's multi-ethnic society, where all citizens live in equality and fulfill their aspirations."

The programme was time bound. Hostilities were to cease within 48 hrs of signing i.e. by July 29. A joint observer group was to be established to monitor the cessation of hostilities by July 31. Consequent to the cessation of hostilities, the militants were to surrender all arms.

Sri Lankan army and other security personnel were to be confined to their barracks within 72 hours. Emergency was to be lifted in Eastern and Northern provinces by August 15, 1987.

It was an extremely laudable role, a situation that we had not been able to achieve in our own country. It was very altruistic of us to offer our services to a third country. The keynote, "surrender of all arms by the militants", was not attainable.

General Depinder Singh met Prabhakaran, the LTTE chief and head of Sri Lankan Army and others; he arranged a surrender of arms ceremony at which the LTTE representative placed his pistol on the table and the Sri Lankans declared amnesty.

Assorted weapons were returned but the weapons handed over were nowhere near the stocks held by them. Although the IPKF had been welcomed by both sides, there were problems galore. Mr. Dixit visited Jaffna for discussions with the Army and Prabhakaran on September 23. His demands included early formation of the interim government with a clear majority of LTTE representatives. He wanted the Sinhalas to be uprooted from certain areas. His nominee as the Chief Administrator was not acceptable to the President. In his ambition to be the sole representative of the Tamils, Parbhakaran had shot some seventy of the top leaders of the other parties. He did not want EPRLF and TELO cadres to come into Jaffna. Prabhakaran's instringence and the death of Thileepan, an idealistic leader who had fasted unto death at a temple on September 26 against the accord, led to Sinhalese–Tamil clashes.

Soon after on October 5, seventeen militants carrying arms were caught by the Sri Lankan Navy. They were brought to the Air Force base at Palaly where IPKF and Sri Lankan Army were based. As they were being taken to Colombo, they swallowed cyanide capsules; only a few could be saved. In retaliation, LTTE killed eight Sri Lakan soldiers and dumped their bodies at a bus stand. Then on October 6, five or six Indian soldiers on their way back from Jaffna city were killed.

General Sunderji arrived and ordered the launch of "Operation Pawan" on October 6. It was now an open war against the LTTE. Their arms surrender was fictional. They committed vile and horrendous crimes against the other militants, groups, and civilians accused of giving information. One such horrid punishment was to make the victim wear a necklace of rubber tyres and then light the garland. A truly barbarous and vile practice, it prevented our troops getting information on the LTTE.

Major Ramaswamy wins PVC

On the night of November 25, 1987, Major Ramaswamy, a Company Commander of the 8 Mahar, who was returning from a cordon and search operation in Jaffna province, drew fire from a temple grove area but he decided to wait till dawn.

The search started at 0530hrs but they found nothing and decided to move back.

Soon after they were again fired at from the temple. So he told Capt. Sharma to hold the enemy with fire while he with his party went behind the coconut grove.

On seeing the enemy they charged them. During a hand to hand fight he was shot in the chest. Despite his grievous wound he snatched the AK-47 from the militant and killed him. Further, he kept on giving directions to his men till he died.

Capture Of Jaffna

The operations by the IPKF had a bad start, a platoon of the Para Commandos, dropped near the LTTE HQ at the University, could not match up with 72 Brigade.

Similarly, a platoon of the Sikh Light Infantry, flown in by helicopters, failed to link up and suffered casualties.

Finally, 115 Brigade, commanded by Brigadier[1] Samay Ram, advanced from the Palaly Air base with the 5/1 Gurkhas supported by 6 Guards and later 4 Paras, and succeeded in capturing Jaffna on the night of November 22/23 after a very tough battle. The administration of Jaffna Town, was taken over by Brigadier Kahlon. He was moved there whilst attending a course at the National Defence College. Everyone talked highly of the orderly way he ran the place. Sappers restored the essential services, water, power, transportation, including tram services. Banks, schools, and hospitals were reopened. Help was provided for the reconstruction of roads, temples and other buildings.

After the initial setbacks, the strength of IPKF was increased. Besides the 54 Divisions at Jaffna, 36, 4 and 57 Divisions plus 55 and 340 Brigades were inducted by the end of 1987.

Almost five divisions with more than normal infantry battalions and a fairly large number of sappers to deal with the I.E.D.'s (Improvised Explosive Devices) were employed. Our troops fought in extremely unfavourable conditions, unfamiliar of the country, its people, with hardly any intelligence feed back, on the LTTE. Without knowledge of the topography of Jaffna, and literally no maps, our men got lost in the Jaffna streets.

The Survey of India being too busy, printing section of the College of Military Engineering, Pune, helped with a few maps.

Our troops with their hands tied behind their backs suffered unnecessary casualties—1200 killed, about four times the number wounded, of which 25 percent lost limbs.

The injuries to our troops by the I.E.D.'s were too high and the equipment available to our Bomb Disposal units archaic. The use of remote controlled vehicles was suggested. Our Finance said they had no money, expenditure up to September 30 was 130 crores! The Defence Research and Development Organisation said that they were not asked to develop such devices. The Engineers did import one or two but there was no concerted effort. When at a U.S.I. lecture, I suggested that perhaps the Sri

Lankan Government be asked to pay for the expensive anti-bomb devices the reply was—all costs are to our account!

The LTTE armed with A-K 47s, carrying walkie-talkies, having intimate knowledge of the country were a formidable enemy. To fight them on the ground of their choosing was a major undertaking, yet there was no military planning for launching such a complicated operation in a country unknown to the troops. The ethnic problem had been festering for years. These had worsened after four years, and yet we went in without knowing that the LTTE was better armed, and had more sophisticated communications equipment than us.

Rajiv Gandhi had a "knee jerk"operating style; he lost in the next elections. When he was out of office, I asked him, as to why he did not have "Think Tanks" for evaluation, and planning. He had no satisfactory answer, except lack of time.

The political side of the picture with various parts of the Indian Government worked at cross purposes. The attitude of the Tamil Nadu government is too complicated to be discussed in this story. The military received little help or guidance in their difficult task.

There was hardly any interaction with the commanders in the field—they were left to look after themselves. Senior visiting officers had very few pearls of wisdom to drop. Gen. Sardeshpande said, "Visits provided little worthwhile guidance, and clarified little as to what the game plan was about".

During a visit, Gen. V.N. Sharma, who took over as the Army Chief on May 1, 1988, put forward the idea of having a CVF (Civilian Volunteer Force) to assist the Army. It did not find favour with our Home Ministry, guardians of our security environment.

The morale of the force remained high. What the troops on the ground felt is illustrated in a glimpse of peace keeping at Trincomalee, related below:

Major D. of an Engineer regiment wrote of an incident on Sept 18, 1987, at the harbour town of Trincomalee-"LTTE, engineered an ethnic clash, used women and children to take part in '*RASTA ROKO* 'agitation. The Sinhalese vehemently objected. This led to Tamils and Sinhalese both resorting to brick batting. A party of the Field Company was already at Trinco, which the LTTE had confined in the Town Hall.

The O.C, Col. Virendra, reached the trouble spot, and in anticipation, contacted me on the wireless to bring reinforcements from another unit. En route some highly agitated Sinhalese blocked my jeep. They wanted me to see a house ransacked by the LTTE, they literally herded me inside it, the Colonel watching me from a far, rushed and pulled me out of the house as our instructions were not to enter a house for fear of being blackmailed. Finally, as we reached the site of the brick batting, we found ourselves totally unprepared for this ugly situation. Could these two-footed infernos be controlled without firing a shot? The Deputy Divisional Commander had also rushed to the spot. We were trapped between two rival groups, trying with agility to avoid being hit by the brickbats from both groups. Luckily, no one hurled a brick at us. At last, a Colonel of the Sri Lankan Army pacified the Sinhalese. We, in turn, pacified the Tamils. They had come well prepared with refreshments and offered us cold drinks. As we sipped our drinks, someone from one of the nearby buildings flung four grenades in our midst. Col. Vireandra at once saw the deadly death high in the air and shouted at the top of his voice "GRENADES." We vanished from the spot and were saved. Three Tamils and a Sinhala were killed —the CO's Jeep's windshield was smashed!

Next month again at Trinco, someone decided to cut off the power supply to the town. The engineer Field Park sent a generator to the Town Hall, suddenly some Buddhist monks flung themselves on the road and would not let the generator pass, as we pleaded and argued we were ambushed. A sapper was hit. We returned fire immediately and a monk was killed in the crossfire. The Colonel on receiving our wireless message rushed with a party and immediately carried out a cordon and search operation. His quick action yielded results. We found 14 Sri Lankan policemen in uniform, hiding under the cots, in almirahs and behind doors, armed with self-loading rifles. On smelling them, we found that these weapons had been fired only minutes earlier. They "Our Friends" had wounded one of our men, it was with greatest difficulty we were able to control our anger. We apprehended them and took them to the Regimental HQ."

What did the sacrifice of so many Indian servicemen achieve? They managed to supervise elections to the Provincial Councils and parliamentary elections in Sri Lanka. Whereas about 60 % of people in Jaffna voted despite ban by the LTTE, there was a 85% turn out at Trincomlee. President Jayewardene and his wife having decided to retire, both the candidates -Premadasa and Mrs. Bandaranaike fighting the election for the Presidentship— left no doubt as to the attitude of the Sri Lankans. They wanted the IPKF out of their country, and had meanwhile palled up with the LTTE.

We stuck on till the election of the President of Sri Lanka. It was a case of "Love's Labour Lost."

The winding up of the IPKF and their orderly withdrawal in a hostile environment was a difficult job, well done. I am grateful to my neighbour who was then the Army Chief, for patiently explaining, how the task was handled. Reducing numbers of troops, living in hostile country poses problems. Here even the Tamils of our own country were threatening to turn against us. It needed very careful planning. The Navy and the Air Force co-operated fully. Perhaps, it is not common knowledge that the Aircraft carrier *INS Virat* and several other ships lay off the coast, ready to intervene in an emergency. Lieutenant General, Kalkat, the Force Commander, and Mr. Mehrotra our High Commissioner, used their charm to ensure that the Sinhalese authorities arranged a proper send off and that the people of Madras –now Chennai –did not come out with black flags. Some facts have been published in Rajiv Foundation papers. These need to be studied.

I recall listening to a very senior and experienced officer of the Ministry of External Affairs, on the reasons that guided us into putting our nose into the affair. The reasons put forward included: to counter the trend by Sri Lanka to involve other nations, prevent the U.S.A and Britain being offered bases, and the Voice of America getting a foothold. Pakistan and Israel supplying arms and ammunitions to influence and control the possible riots to throw out the Tamils.

Finally as we all know, Israel played a major part; it not only supplied arms and ammunition it also provided expert advice. The Conflict between the President and Prime Minister of Sri Lanka on settling Tamil problems bedevils all efforts. The later has taken a rigid pro-Sinhale stand and has demanded surrender of all LTTE cadres with weapons, prior to any settlement. This is not in the best interest of their country.

Now we are glad that through the initiative of our Norwegian friends an accord has been reached, and hope better sense will prevail.

Note

1. Major General Samay Ram UYSM,AVSM,VSM

❮14❯

Buddha Smiles

Our *Gurus* (seers), members of the Reforms Commission, enjoined on us, the inheritors, that a country like ours wanting to invest the maximum resources in economic and social development and desiring peace, must also take note of the threat of nuclear weapons.

This chapter is in response to this policy and the threats by our neighbour, Pakistan.

It was on September 23, 1997, the Pakistan Army day, that Mr. Nawaz Sharif, the Prime Minister, first announced Pakistan's nuclear capability. He again reiterated it on April 9, 1998, on his way to celebrate Eid, at Lahore. He, also on that holy day, announced the test firing of the *Ghauri*, a 1,500 km range missile, that can carry a payload of 700 kg. He claimed that it would help bolster the defence of the country.

The Government of India decided to test its Atomic capability. The greatest event of May in the year 1998, was Pokhran II when India declared itself an Atomic Power. It was a more scientific repeat of our Pokhran I in 1974.

Five tests were held; a very high degree of sophistication was achieved. The greatest gains were ability to use smaller mass as required in artillery shells. Also nuclear implosion ability for making the more powerful hydrogen bombs.

The tests of May 11 and 13 starting on the Buddha Jayanti day, were the culmination of the undertaking of the B.J.P to go nuclear. It caused such a storm in the country that it will take quite some time for the dust to settle down. Our society with 17 major parties in the Parliament, and nine in the cobbled up Government, is too disparate to discuss such sensitive matters. The Government instead of being congratulated for having kept the event secret, even from the C.I.A, was dragged over the coals for not having taken the opposition into confidence. Some leaders with their sole interest in gaining power must be joking when they talked of prior consultation.

Hardly had the euphoria after the tests died, when Pakistan, too, staged a series of tests—one more than India. This was one upmanship. The possession of the Atomic weapon has led to a great show off in Pakistan.

Pakistan gives undue importance to the Bomb. Munir Akram, their Ambassador to the U.N, proudly announced it's nuclear capability to the General Assembly.

The fact is publicized all over Pakistan—a lighted model of the Chagai Hills, the Pakistan test site in the western hills of Baluchistan, is displayed at the entrance to the capital at Islamabad. The day of the test has been glorified and is celebrated throughout the country. Models depicting atomic explosion are displayed in the parks of its cities.

What ails Pakistan is what my mother used to say, "*Nai Nai Mussalman—Allah Allah Pukare*" (A newly converted Muslim lady keeps on repeating Allah-Allah).

To get the bomb—Bhutto, the Prime Minster had declared "We will go nuclear, even if it is necessary to eat grass." The urge was so overriding that Nawaz Sharif—the last civilian P.M.—refused a US offer[1] of a $5 billion package for showing restraint made by Deputy Secretary Talbot in 1998.

The refusal of this offer shows the great value attached to this acquisition. Pakistan intent on declaring its "Bomb" went ahead with its test on May 30.

Within three weeks of the test the value of its rupee fell from 54 to 40 to the dollar.

Pakistan is searching for its nuclear doctrine whereas India, like China has declared a "No first use" policy. Pakistan, which once upon a time asked for a nuclear free zone, has an agreement with India not to attack each other's nuclear establishments. This list is exchanged dutifully every year. They now appear to be so obsessed with the new toy that they want to use tactical atomic bombs.

The United States imposed sanctions, which gave another handle to the politicians in the opposition. They started a virulent campaign, "Why didn't the Government think of the economic consequences?" Luckily, even the illiterate villager understands that "National Security" cannot be priced. The build up of our compulsions for holding Pokhran II tests were overwhelming. In 1996, we as the 62nd country, ratified the chemical warfare convention, and gave up our option of a deterrent. Our Government's resolve to have a viable "deterrent" was stressed soon after by P.M Inder Gujral. He was praised in the "Times of India" editorial of July 1, 1997. Interestingly, it was under the apt quotation from J. F. Kennedy, "*Let us never negotiate out of fear, but let us never fear to negotiate.*"

Keeping the above in mind, even half wits, would laugh at the canard "How can a country, of destitute people, afford the expense of an Atomic Bomb." The most vocal is a party, with an ideology from a country, with a huge stockpile of bombs.

About costs, let me explain, our research on production of electricity from atomic power, gave us the knowledge. Decades of work by our Sappers units, prepared the shafts and chambers at Pokhran. Nothing was imported—possibly that is what irks some Netas—there were no kickbacks.

The people forget that we, with a population of nearly a billion have the world's second largest human resource. Consisting of the vast peasantry, a thriving industrial sector of over 150 million, who produce most of the goods and services required for our economy. Travelling on any highway, it strikes one forcefully, that almost all the transport, tractors, and machines seen in the countryside are of local manufacture. Besides the two old categories of workers we now have a growing, third generation of people, working on their home P.C's, satellite dishes, signal splitters, software for export and what not. Their contribution in the information technology, is being accorded incentives which shall no doubt lead to a "roller effect." There is a growing demand for Indian contribution. If only we could overcome the present fissiparous movements—in the name of ethno-religious differences—which have been the bane of our country, we can see a bright future. The pulling down game has extended to such an extent that even the pillars of state—the Judiciary, the Civil Service, Defence Forces and the Police—are threatened.

We are lucky that we have 500,0000 tons of Thorium reserves in the form of Monzonite sands from which we process through trisodium phosphate-thorium salt to the metal. Our fast breader reactors in operation since 1983, produce more Plutonium, than they consume. We, thus, have the easiest way to fission bombs. Only about 5 kg of PU is sufficient for making a Nagasaki type bomb.

As the only Indian to have taken part in a NATO atomic exercise with troops in Germany in the year 1954, I had a window on the European/U.S approach to Atomic Warfare, denied to the hundreds of theoretical writers.

It was one of the largest exercises ever staged and aptly called "BATTLE ROYAL." Troops from seven nations took part. America for the first time displayed its 280 mm Atomic Guns. With barrels as long as the old 6-inch coastal battery guns these were towed by huge tractors. The one I saw was parked on the roadside. These were difficult to hide and camouflage. Now, Atomic shells can be fired from even smaller caliber guns like our Bofors 155mm-Medium guns. The commanders had a choice of weapons; they could use "Tactical" bombs of various tonnages 1-10 ton capacity, also the new Neutron bombs. The Neutron bomb, also called the capitalist weapon, emitted a much higher larger dose of radioactivity, but caused lesser damage. These were to be used against the Russian tanks forming up for attack-called F.O.F.A—for a breakthrough NATO defences. The idea was that the massed tanks, would be subjected to massive radiation which would even penetrate tanks in hull down position.

Yet, the damage to the surrounding areas would be limited, which would not hamper the movement of own counter-attacking force.

The commanders of the opposing forces were allotted an array of these weapons and were free to choose their targets; the only restriction imposed was that before firing the weapon the commander shall obtain political clearance.

The battle raged night and day for three days—Tanks and troops of the seven nations maneuvered over thousands of kilometers. Highly decorated, and keen commanders vied with each other in the selection of many a suitable targets but believe it or not—not one Atomic warhead was fired. The final conference was a very stormy at the end of the exercise. It was a waste of time said the irate till a Lt. Col. of the U.S Army got up and said, "Gentlemen—now please imagine yourselves in the chair of the President of USA. If you allow one atomic shell to be fired—you face the whole of New York being flattened in retaliation." There were no takers.

Pakistan talk of "Tactical bombs" was ruled out at the "Battle Royal" exercise. It appears that Sarfraz Khan, the Pakistan Military Adviser, who also attended the NATO exercise "Battle Royal", has not been introduced to General Musharraf. The General persists in his assessment that India was prevented from taking any action against Pakistan because of his message of warning sent to Delhi with every dignitary calling on him at Islamabad.

Today, General Musharaff is under great pressure from his people. They accuse him of having sold Pakistan too cheaply to the Americans when compared to the comparatively small payment of $891[2] million for its co-operation in Afghanistan. According to Mr Shaukat Aziz, the Finance Minister, told it cost Pakistan $2 billion. On the other hand, it received $600 million as cash grant and is expecting 5 billion dollors in aid shortly from the U.S. In addition, debts amounting to $ 12.5 billion under Paris club are being rescheduled. A $ 62.2 million debt from Denmark, the Netherlands and Commonwealth Development Corporation of the U.K. has been written off.

One of our scientists went to the press asking that the "Neutron Bomb" be developed. But there is no mention of the Hydrogen Bomb.

Pokhran II tests of May, 1998, included tests of smaller war heads, these are called "Tactical Warheads" as well as a fission device, the basic requirement of the Hydrogen bomb.

Now we have decided on using this weapon for the Hydrogen bomb.

The Hydrogen bomb is not all that secret a device; it is described in many books.

The fact is that we tested it and that we have ample stocks of Plutonium, which forms "fission trigger" of the H-bombs; details are now public. The "Trigger" is in the shape of a soccar ball, less than 12 inches across. Around it is an inch thick cover with the detonators. The centre trigger or primary stage is filled with Plutonium along with a small amount of tritium and deuterium, the two forms of heavy hydrogen in gaseous form.

The primary stage is sufficient to level a small city by itself, but gives the H-bomb the energy to ignite the much more powerful secondary stage. This is litium-6 deutride, which manufactures its own tritium from solid Lithium–6. Surely we can manufacture Li-6 and have the manufacturing details of the Bomb. A recent disclosure in the press that Pakistan has a better nuclear arsenal should not cause us any alarm.

General Musharraf claims that his messages to India through visitors stopped the escalation during Parakaram. Pity is that the novice in the game has not been warned .

Pakistan by threatening to use the Bomb against us in a conflict risks mutual destruction. In this connection the Chinese thinking handed down from Sun Tzu-—the importance of, deception—incorporated in Mao's military thinking is important. The danger here is that unless we can threaten to deliver massive retribution in quick time.

The F-16A/B's of the Pakistan Air Force, Range 1600 km/ based at Sargodha are capable of delivering "A" Bomb. They also have the Chinese M-11 and the Pakistani variant Haft 3. Further, there are the North Korean supplied Nodong missiles called:

Ghouri I (Haft =5) Range—1300-1500 km
Ghouri II(Haft =6) Range—2,000-2,300 km
Ghouri III Range—3.000 km—Tested Aug 2000
Its stock estimates vary from 15-70.

From the analysis of the reports of Pakistani tests on May 27, ordered by Nawaz Sharif, on the advice of General Karamat, the Army chief, to bolster the morale of its soldiers–it is clear that the first tests proved the technology to manufacture first generation fission bombs. The two tests on May 30, indicate testing of sub critical, heads for tactical bombs, possibly even–for artillery shells. In this connection, perhaps it is not generally known that Britain does not manufacture these–it uses American shells.

The two Indian tests held later, related to "Mini" sub Kiloton Nukes. These are used as warheads on missiles and Artillery ammunition. Today, atomic shells can be fired from guns of caliber, as small as the 155 mm (Bofors). For this gun, ammunition type W-82 is available in sub kiloton and 4-5 kilotons yields.

Cohen in his book "*India*" quotes D.K.Palit calling Indo-Pak conflicts as "communal riots with armor." The Atom Bomb now hogs the limelight. Pakistan feels that, its possession, gives it ability to wrest Kashmir from India. In any case by threatening to use the bomb, they have internationalized the dispute.

Knowledge of the heavy elements–Thorium and Uranium is as old as, the hills. Hundreds of books describe development of the Atomic Bomb. Thus the ability to explode a nuclear device does not involve an "*Ilham*"(A revelation by God Almighty). Yet, the Genie once released is a problem—the control of the radioactivity–even in the spent fuels is a tremendous problem: the implications of which have been very forcefully or perhaps too vividly brought out in the National Geographic Magazine of July, 2002.

The picture is glum, as the half-life of Plutonium, PU-239, is 240,000 years. But it should not deter countries like Pakistan and India –short in energy resources from using atomic power. In a truly sensational media manner, the dangers are exaggerated. It is only at the end that one learns that the Cesium dosage detected in Essic, the photographer, was due to reindeer meat eaten by him in Sweden, and it will gradually disappear without causing harm.

It is time, that someone spoke objectively about Pakistan Atomic Power—their great, scientific achievement and assessed its likely repercussions.

Will Pakistan use the Doomsday weapons in the event of a conflict? No one can predict, but there is every danger in their using some weapon of surprise for example, they could use a tactical bomb to disrupt communication in Kashmir or destroy our co-headquarters to attain their goal of 'liberating' Kashmir.

We should, clearly, tell any hotheads who wish to indulge in such an experiment that there will be massive retaliation by us.

In case of a chemical-biological attack on us, we have already said that we will be free to use atomic weapons. In the case of a nuclear attack on us—even by a tactical bomb, we would be free to attack their nuclear establishments. One bomb on their Plutonium stations which is located on the north bank of the Jhelum River, and another on Kahuta will be the death knee of all the Pakistan Atomic Power. The bomb on the Jhelum River is an easy target, it would not only cause tremendous destruction the release of Plutonium into the waters of river Jhelum will poison the water and be a menace to the whole of the people living the downstream and possibly also the people living on Ichhogil canal, the headwaters of which start at Jhelum. (The Ichhogil canal, built under the Indus Water Treaty, takes waters of Jhelum under the river Ravi to the river Satluj.)

Another fact is that our chemical industry produces as an intermediate a million tons of ethylene oxide, a very powerful explosive. We also have methane, butane, propane, propylene oxide and propadiene, when dispersed from canisters to form aerosols and detonated, explode devastatingly and have a much greater overpressure than an atomic bomb at ground zero. Lahore, if attacked with FAEs can be destroyed instantly.

Notes

1. *South Asia On A Short Fuze*—by Praful Bidwai and Achin Vinaik, published by Oxford University Press.
2. *The Hindu*-March 9th, 2003.

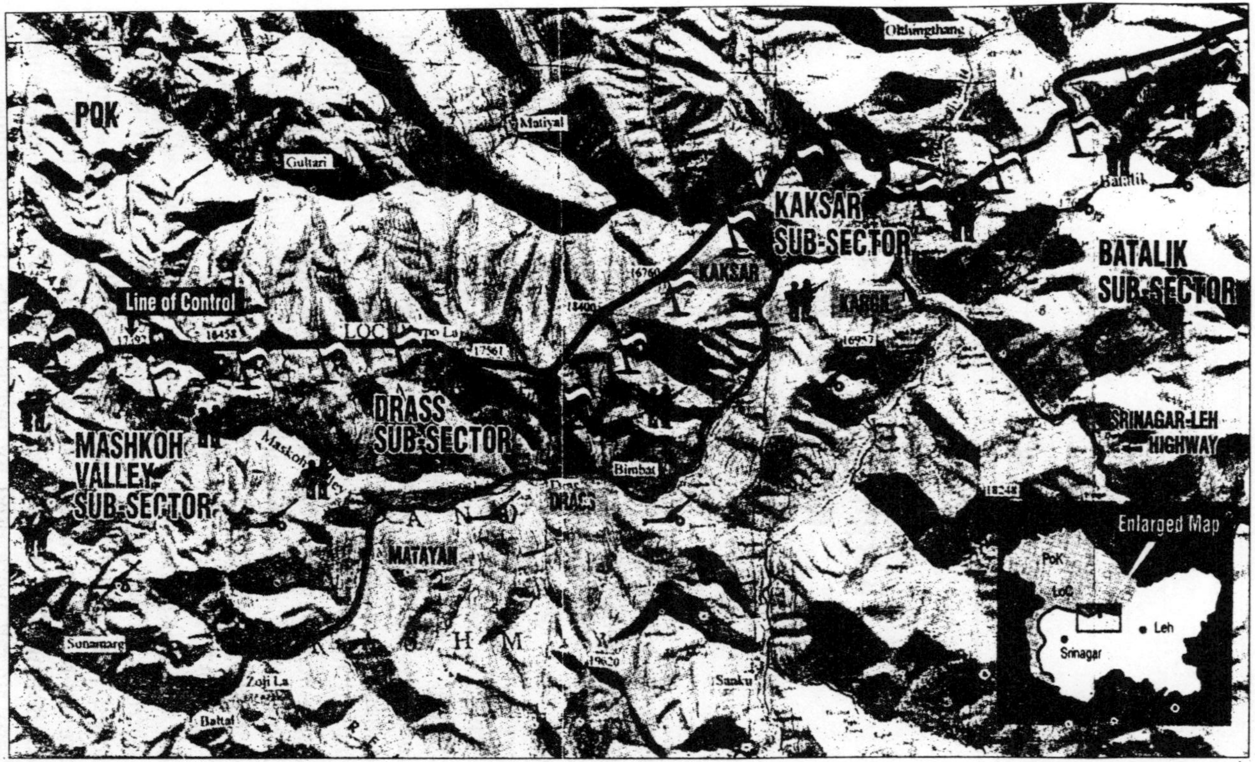

Kargil Sub - Sectors 1999

(Map not to scale)

‹15›

Kargil and Parakram Recollections

Manoj Joshi, whom I met, published an eyeopener in the '*Times of India*', based on a report by the C.G.A (Controller General of Accounts), saying that" *Kargil need not have happened,*" if we had a smarter Defence management. His story is summarized below:

The IAF ran Flights 181 and 182, these were top-secret missions flown daily. Flying parallel to the border, well inside our territory, these flights one from Leh and the other from Jamnager would fly towards each other. Enroute they would pick up radio and radar emissions from radio sets or mobile surface to air missile batteries. We thus built up a "Library of each emitter" a most important means of tracking movements of Pakistani troops.

In 1987, 12 years before Kargil, the Air Force reported that the Avro's used for the flights need to be replaced and the electronic snooping devices imported at a cost of Rs. 30 crores. The DRDO said it will develop and make the electronic receiver-recorders. It took six years for the Ministry to find two old Boeing 737 aircraft as replacement. Then for two years, these aircraft were used for ferrying VIPs on their jaunts—the excuse was that there was no hanger space at the testing establishment at Bangalore. For mounting the electronic system, these aircrafts were only given to BEL in July 1995, and January 1998. This lapse of system which would have provided early warning, points to a failure of intelligence.

Possibly, it has been mentioned in K. Subramanium's report on Kargil.

I also recall reading the captured diary of a Pakistani officer of 1999 from which it is clear that the various posts visited by him were well established well before his recorded journeys from January 1, in the New Year (1999). What is surprising is that the officer talks of test firing the weapons and exchanges of signals. Surely, our garrisons should not have been ignorant of such activities.

It makes one accept the uncharitable remark that they were blissfully escorting fair visitors to Leh.

Jasjit Singh called Kargil, Pakistan's fourth war for capturing Kashmir, but that is an exaggeration. The press is divided as to whether Kargil was a war, limited war, battle or a conflict. The term conflict would obviously be inappropriate—some 500 braves did not return home. The scale of operations qualifies it as a battle. It cannot be called a war; the Foreign Minister of Pakistan came to Delhi for a discussion.

Instead of raising public support, our politicians vied with each other in their one upmanship game. Apart from liberal donations for PM's National fund, there was hardly any reaction by the people. There were no women's circles knitting socks and gloves for the Jawans. It was business as usual for the people.

Brigadier Surinder Singh commanding 121 Infantry Brigade received the first report of occupation of the heights above Kargil from a shepherd on May 3. Reconnaissance patrols were sent out on May 6. These suffered casualties. On May 9, Pakistan artillery scored a direct hit on the ammunition dump at Kargil. On that day, the Chief of the Army Staff, General Malik left on a goodwill mission to Eastern Europe, nothing unusual was suspected.

Extensive patrolling followed. On finding large-scale intrusions by the enemy, things took a serious turn, additional troops from

8 Mountain Division were inducted, and the decision was taken to recapture the lost posts.

The Cabinet Committee on Security met on May 25, and taking note of the situation, it approved air strikes against the enemy, within our own territory, without infringement of LoC (Line of Control).

Air Force ground attack aircraft mounted their first strikes next day. The priority targets were the Tiger Hill and the post in Tololing sector.

May 27, was a black day for India. We lost an MIG-21 flown by Squadron Leader Ahuja. It was shot down with a Stringer missile. Another MIG-27, crash-landed across the LoC due to mechanical failure. A day later, an M.I-17 helicopter fitted with rocket pods was shot down in the Drass sub sector, it tragically led to the death of its crew of four. Our Air Force was obliged to change its tactics.

The tenacity and bravery of the officers and men of our famous Rajputana Rifles won the day, they retook Tololing heights on June 13.

The 13 Jammu and Kashmir Rifles were ordered to support 18 Grenadiers to capture the "Hump", which infact is an extension of the Tololing feature. The attack took place on June 17. Captain Batra of the J&K Rifles personally led the attack. Captain Jamwal also of 13 J&K advanced for the kill.The enemy held 7 bunkers on this `15,000 feet high Pt 5140—fired illuminating shells. The situation was difficult till the C.O Lt. Col. Joshi personally fired two FAGGOT missiles which silenced the enemy.

The Hump was captured, unfortunately Batra who was awarded a PVC was killed. Rifleman Sanjay Kumar, one of his Jawans, was also decorated for his bravery in the battle for point 4875 in the Mushkoh valley.

The battle of Tiger Hill started on the night of July 3/4. The 13 Grenadiers supported by the 3 Sikh covered themselves with glory, Grenadier Yoginder Singh Yadav attacked the three bunkers with the help of his comrades. He lobbed hand grenades to kill the enemy—earned and lived to wear a Param Vir Chakra.

Four PVCs were earned—two by officers. The other officer, Lieut. Manoj Kumar Panday of the Gurkhas also won it posthumously.

Three infantry weapons available were the old 65 mm—produced by ordnance factories. Konkors Anti-Tank Missiles produced by BDL and the Russian 30 mm AGS-30 automatic grenade launcher, a deadly weapon with a range of 1700 meters, which had been specially chosen by General Bammi for Siachen. Two major deficiencies were in UAV's for information and target indication, and Gun Locating Radars. The operations lasted some 55 days. Diplomacy too played its part. After the recapture of Tololing, the Foreign Minister of Pakistan came to Delhi to discuss the matter with our Minister of External Affairs, Jaswant Singh. No headway could, however, be made as Pakistan struck to its stand that its Army was not involved in the fight, which was fully in the hands of the Mujahideens.

India brought the bodies of two Pakistani officers and tried to hand them over to the Pakistan High Commission at Delhi. There was initial hesitation, but finally these were accepted. They, however, refused to accept the involvement of the regular NLI battalions and SSG Commandos and attributed all the military operations to "Kashmiri Freedom Fighters." There was a lot of blood letting. There was evidence of mutilation of the bodies of six of our Jat Jawans, and shooting in cold blood of one of our pilots. But Pakistan denied its involvement in all these gruesome acts.

Our infantrymen courageously advanced on July 2, in the Batalik area, a Naik of 5 NLI (Pakistan's Northern Light Infantry) was captured. It proved beyond doubt that Pakistan had intruded on our side of the LoC, if proof was needed.

Politically, the end came when Nawaz Sharif, Pakistan's Prime Minister, made a special request for an unprecedented meeting with the American President, Bill Clinton on July 4, the national holiday. As a result, Pakistan complied with the US President's request and withdrew its forces from the LoC. Incidentally, Nawaz Sharif was received at the airport by the Saudi Ambassador who primed him on the wish of the President.

The Director Generals Military Operations (DGMOs) of India and Pakistan met at Attari on July 11. Pakistani troop withdrawals were announced on that day, but fighting was reported by the media till July 23. The end of hostilities was finally announced on July 26.

This was the first ever war to be brought to our homes, by the television. The media was free to travel to Kargil without restrictions up to June 4. India's first war in the age of the Internet and Television. Lady reporters like Barkha Dutt, wearing a steel helmet reporting from the front, brought pictures of war to our drawing rooms. Unlike papers and T.V, the Internet being an interactive medium a number of people visited the web-sites to get information from the battlefront.

It was an Infantry-Air Force battle; unfortunately, neither of them were properly equipped for the task. Looking at the photographs of our troops, two old war veterans noted that the infantrymen were wearing non regulation ammunition pouches-popularly known as "*Chini*" these were preferred to heavy regulation issue harness produced by the ordnance. They were either taken off from the killed militants, or copied by unit cobblers using canvas from ground sheets and chaugals. Men wore black scarfs, *"Patkas,"* bought from regimental funds. This lapse points to a major defect in our equipment development system. It would appear that the DRDO, which is interested in expensive equipments, has little time for the poor infantrymen. Would it not be better if the development of the dress and the equipment of the solider was entrusted to the Infantry school.

Most photos showed a number of men wearing camouflaged cotton/synthetic mix trousers and worn out footwear—surely they deserved better. This shortage was despite the fact that we now have five ordnance factories that produce equipment and clothing on cost plus basis. We continue to run these outdated factories. Thirty years ago, the Administrative Reforms Commission had recommended the closing down of the ancient Shahjehanpur tailoring factory. Instead, today we have two or more clothing factories.

There was even a talk of shortage of snow boots, and how it was difficult to import these as Pakistan had already cornered the supplies. This was complete nonsense. For decades, I have seen snow boots being manufactured in India. Our specifications unfortunately insist on specified weight. No attempt has been made to design a lighter boot.

Finally, a word about the use of air powers. Fast fighters are of little value for "Fire Support" in the mountains. We need armoured Gun Ships.We used smart bombs, laser guided bombs. Strap on guidance systems—costing 21,000 dollars—have been in the market for years. We could also use Napalm bombs. Ideally, the Air Force requires relatively slower moving armoured ground strike aircraft, like the American, A-10, or SU-25. These will require special provisioning.

The Bofors guns were seen firing from open slopes. Without these howitzers capable of firing at steep angles we would not have been able to provide our troops with fire support. Our gunners also used 130 mm guns designed to fire at a maximum elevation of 45 degrees. Though the media showered praise on the 155 mm Bofors Guns, the Russian 130-mm medium guns and the 105-mm Indian Field Guns and the Heavy Mortars made a major contribution. Five regiments of the artillery equipped with the Indian Field Artillery guns—designed and produced in India—provided the main close support. Out of the three Artillery Regiments awarded unit citation, two of them—141 and 197—were Field Artillery Regiments.

Rocket Regiments, capable of delivering 132mm rockets in 20 seconds, were used for area neutralization. 2,50,000 shells, Heavy Mortar Bombs were used during the operation of which 9,000 were fired the day the Tiger Hill was regained.

Some 800,000—155-mm Bofors rounds were fired, each was imported at a cost of—US 1000 dollars—nearly worth half a lakh each—it was a costly war. The gunners lived unto their motto "*Izzat-o-Iqbal.*" This has been recorded in a book brought out by the Regiment of Artillery. Whilst our gunners worked from exposed locations, a photograph in "*India Today*" June 28 showed a Pakistani 5.5 inch Medium Gun, in an emplacement built with Bailey bridge panels and its crew, living in comfort, was immaculately dressed in light coloured fatigues.

Having dealt with Japanese bunkers in Burma, let me tell you that such emplacements could only be destroyed with guided missiles. Wind proof warm clothing, bulletproof vests and devices like GPS—Hand held position finders—and infra red viewers, and RPV's were required not only for now, but as insurance from being surprised in the future. All these can be produced; what we need is an honest and competent procurement system. It is believed that a new set up is in process of being installed.

A lot of heat has been generated on the import of light aluminum alloy coffins at high prices. These arrived too late to be used. The Army managed with locally made wooden caskets for bodies of the killed to their homes. Another scandal is the import of sleeping bags from France, surely these can be made locally.

Will it be an another case of decades of investigation like that of the import of Bofors guns? A swindle. This gun was demonstrated at Karlskoga, Sweden, to Colonel Sharma and I, and offered to India for manufacture, in 1959. Twenty-seven years later, in 1986, 410 guns were imported in haste–as a hush hush weapon. For many years, urgency was forgotten. The bureaucrats went on a trip to Europe in June of the previous year, to look at guns. A Chief of the Army staff jumped into the arena. He advised the cancellation of the manufacturing contract. It was accepted. The result has been that we have been importing every round of the 155 mm Bofors ammunition at a cost of a thousand dollar each, instead of indigenously manufacturing. Imports are from South Africa which had only two ordnance factories in 1947, compared to 17 in India. Their level of technology and manufacturing was at about the same level.

The basic question as to how and why the Defence Secretary decided to import on the advice of the Scientific Advisor has never been answered. Meanwhile, we are throwing good money after that.

All papers talked of the accuracy of Pakistani shelling. There were reports of presence of "Flying objects," which baffled the Army." I recollect a weekly magazine had published the orbit of the opposing forces—Pakistan had some 90 Flying machines!

A great amount of publicity was given to the request of Brigadier Surinder Singh, Commander of 121 Brigade, for satellite pictures, RPVs and Gun locating Radars.

It almost turned into a parody of shooting the messenger, and not heeding the message. It brought to forefront the extent of ignorance of the general public on military matters.

We are proud of our space scientists who have successfully launched the GSLV, and earth observation satellites, the quality of their imageries is of a high order but 30-feet resolution does not meet the Brigadier's request for detailed high-resolution photographs from satellites.

To fight battles successfully, one must know whereabouts of one's enemy. Remotely piloted vehicles (RPVs) have been in use since 1982 for reconnaissance by Israel. The Kargil garrison required RPVs. These radio controlled toy planes are neither expensive nor difficult to manufacture. These are flown from small airstrips some distance behind the front line, or can be launched from a vehicle. Properly prepared for the flight they rise steeply and follow a pre-determined path, and send back video information and photographs of the terrain. In case of failure, the computer on board guides the plane to land back near to from where it took off.

Our Defence Research and Development Organisation (DRDO), has been producing prototypes. A model called "Falcon" was test flown in 1955. A similar device, a target plane, is already in service. Delegations have visited Israel and other countries to look at different models. Samples were imported and tested. But somehow our procurement procedures did not take any timely action. I learnt that a Pakistani had one produced by their ordnance factories within six months.

These have been in use for information gathering and target designation the world over. Twenty years later, we are still researching and carrying out trials. Our troops should not have to operate blindfolded. Report says that the Israel's RPVs were approved in 1996.

The Brigadier had also asked for Gun Locating Radars. We have been using sound ranging systems for location of enemy guns, from times immemorial. The principle of tracing the flight of a shell, backward along its trajectory is well understood. General S.N. (Tindi) Sharma recollects having been shown a Mortar Locating Radar in 1978 at Devlali. If we had not been able to provide Gun Locating Radars to our troops in 21 years, surely, there has been a major systems failure.

No body dare question the Ministry of Defence. A sample of compliance was "*India Today*" published an interview with the Brigadier's Divisional Commander at which the General is understood to have told the Brigadier, *"By seeking things like radars, remote piloted vehicles, and high resolution satellite imaging, which are not on the army's inventory, you are asking for the moon."*

We have since imported some RPVs from Israel. The latest news is that America has agreed to supply 6 Gun Locating Radars to India at a cost of 146 million dollars.

There have been many attempts to improve the system of provisioning of defence weapons. In this context, the remarks of Dr. Raja Ramanna at a recent United Services Institute seminar, that the entire defence procurement system and Ordnance factories need to be dismantled and reconstructed, are significant.

Why the systems have failed? Our greatest folly is in control of the development and production of weapons and equipment by the Ministry without involving the users, at the grassroot level, and at production factories.

To counter the threat of "Future Kargils," instead of harping on crossing the LoC, let's take some measures short of war. These could include high power microwaves to disrupt the sophisticated communications of the enemy. They were able to pick up signals from our emissions, and bring down artillery fire. The press reported how within half an hour of a reporter taking a TV camera shot of a bunker, Pakistani guns opened up and caused six casualties.

Let us arise, and shed complacency. We have treated the severance of the lifeline to the Buddhist Ladakh, much too casually. Over the years, there has been no sense of alarm. A former Army Chief in a television interview opined that we build another road out of range of Pakistani shells.

The "Defexpo 02," at Delhi, had brought strong contingents of salesmen from world over, supported by their Governments to get orders from the Indian defence market worth Rupees 15,000 crore. The figure in the new budget has been raised to rupees 21,410.63 crore. It is imperative to improve our manufacturing capabilty and help in reviving our industrial recovery. But first we have to promote understanding between the users and the manufacturers.

Ministry of Defence has a vital role. The Defence Minister, on a visit, found that the troops at Siachen were finding it difficult to move in the absence of "Snow mobiles." He obtained their sanction, by asking the concerned officials to visit the locale.

The Government is revamping the procurement system. The organisation for procurement is under a Special Secretary, (see Appendix V). Unlike other countries, our procurement executive does not appear to deal with exports. In the British and the French systems, the chief executive dealing with procurement also looks after exports. He is the single window to deal with the foreign trade. The special secretary in our case is in a better position to offer barter sales and reduce our import bill by offering to supply sub-assemblies and even complete assemblies such as trailers. This method is followed by Australia. There is a market for SAA when we are in a position to produce at commercial prices. We could also export grenades, shells, mines plus uniforms, rubber and electronic goods besides small seacraft.

As far as I am aware, the basic cause of our "System Failure" is that the need for closer understanding between the users, the services, the producers of weapons, DRDO, Factories Inspectors has not been tackled. For promotion of that understanding culture, the Reforms Commission had recommended the establishment of a Defence Industrial College (see Appendix III). The understanding can be promoted by providing technical education at a University level as in America, which will promote an understanding at the intellectual level. If that is not acceptable for some reason, we can promote similar understanding by following the pattern of the Royal Military College of Science. The syllabus and training at this Military College is now run by the Cranfield University.

For the information of my colleagues, paranoid about security, let me disclose that the dean of the University, which guides the Royal College, is a person hailing from Ludhiana. In my book, *Indian Armed Bazaar*, I had written that India was the top importer of conventional weapons in the world and held that position even in 1992. We still import at the drop of a hat. Is it due to a system of collecting commissions?

Every time after an import is announced, there is offer of local prototypes. The LRDE is offering a Gun Locating Radar. The Central Electronic Corporation at Hyderabad says it can manufacture it, next year.

Whereas the Navy has made considerable progress in ship manufacture due to setting up a cadre of battleship constructors –there is a wide chasm between the Army and Airforce users, developers, and producers. The bill for the import of aircraft, without which we cannot exist is very high. An estimate is Rs 6,000 crores for 60 trainers and Rs 12,500 crores for replacement of only 250 of our overaged MIG 21's. The import of these two items alone would use up the entire capital budget of Ministry of Defence. We have to think again. I was told at the highest level that our "Kaveri Engine" for our Advanced Light Fighter will fly. Let's work on it and remove the stigma of being one of the top importers.

Besides aircraft, the import of proven technology items, such as Gun Locating Radars and ammunition for the Bofors guns, must cease.

In conclusion, we laud the tenacity and bravery of the young officers and Jawans who won the day in Kargil. Providence helped us by gift of an early, hot weather so that the Zojila pass was opened three weeks earlier. American intervention helped to bring an early end to the hostilities.

Our brave soldiers overcame all obstacles and succeeded in evicting Pakistani forces, to reoccupy the heights, but at what cost? The 474 officers and men were killed. Four Param Vir Chakras were earned, more than in any previous war.

Parakram

Operation Parakram was a reaction to the terrorist attack on the Parliament on December 13, 2001. It was the exercise in coercive diplomacy to pressurize Pakistan to stop infiltration and attacks by the *Jehadis*. Although a certain amount of equipment had been imported there is no doubt that our adversaries knew that we had not been able to import gun locating radars and the number of UAVs imported, was too small to be effective.

Despite our mobilization, a major brazen attack took place at Kaluchak on May 14, when the militants entered an army camp and shot up women and children. In response to this incident, our Government wisely placed all police and paramilitary forces under the military commanders, a very essential step upholding principle of unity of command.

Despite the threats of the terrorists, the successful and impartial elections in Kashmir were held due to the presence of the increased military forces. To save the hardships on the soldiers, the Government decided to send the troops back to the barracks in October 2002. Soon after, there was another terrorist attack on the temple in Jammu, and the infiltrations continued. To reduce the infiltration, one of the steps necessary is to arm the border police and village guards with suitable weapons. A possible weapon produced by our ordnance factory is a 12 bore pump action gun. This gun firing a #18 shot aimed at the legs is a sure way to stop people. The only problem is that as it is not an imported weapon, this does not find favour with the Border Security Force, although it is used by the Army.

The operation Parakram, besides costing the government some Rs 800 crores, has had a devastating effect on the engineer troops employed on laying the minefields. Some 80 sappers were killed due to defective fuses. Further, casualties have occured during de-mining.

The Politics of Kargil

The Kargil war, whose memory is still fresh in our minds, yet again reminds us of the stupendous efforts and never-say-die spirit of the Indian forces in times of adversity. As the in-house study of the 1999 Kargil war has recently come in, questions are being raised once again on the viability of various decisions being taken by army chiefs and the concerned ministries.

According to the classified six-volume report, the government's delay in granting clearance for use of air power resulted in Pakistan getting sufficient time for completing its build up and consolidating its defences, resulting in increased Indian casualties. The Indian army lost 474 men in the war. The report points out that the target information was sketchy and many a time the coordinates and geographical references were off the mark. Voice communications, too, were unreliable.

Radio intercepts show Pakistanis had knowledge of flight plans of our aircrafts before missions were lauched from Srinagar and Awantipur. Poor communication also led to our Mirage-2000 missions, from Adampur and Ambala, not being identified in time by our own army formation in forward areas, as these aircrafts were maintaining radio silence. Communications from control and reporting centres to electronic warfare radars; from combined operation centres to air defence nodes, remained insecure.

Air space management in the tactical area was shoddy too. The IAF lost a MiG-21, a MiG-27 and one Mi-17 chopper. Such losses were predominantly due to the pilots digressing from briefed sortie profiles. A Canberra bomber was also hit but the pilot managed to land it.

In order to overcome several such loopholes, the study states that to remove red-tapism in decision making the arrangement needs to be institutionalised at the politics-military level. In what is likely to be hotly opposed by the IAF, the report says medium lift attack helicopters should be transferred from the IAF to the army. It states that these were not efficiently exploited due to lack of adequate coordination between the two forces.

The portion in the report on maritime operations suggests the need for the armed forces to have capability to launch a brigade-sized amphibious force with its integral mechanised infantry, artillery and aviation components optimised for its envisaged role.

Despite several pitfalls, the Kargil war proved once again the supremacy of Indian Armed Forces and gave Pakistan a mighty blow. Still numerous questions regarding greater empowerment of army to avoid slackness and delay in operations, remain unanswered. Various decisions regarding many new changes are under consideration. But, only time will tell when the governments will quit the 'blame-game' and work for the betterment before anyone else again takes an advantage.

No war is ever won, its only statistics that tilt in favour of one side on paper. Nothing could be more aptly said than the words of Arthur Kelps:

"Tolerance is the only real test of a civilization."

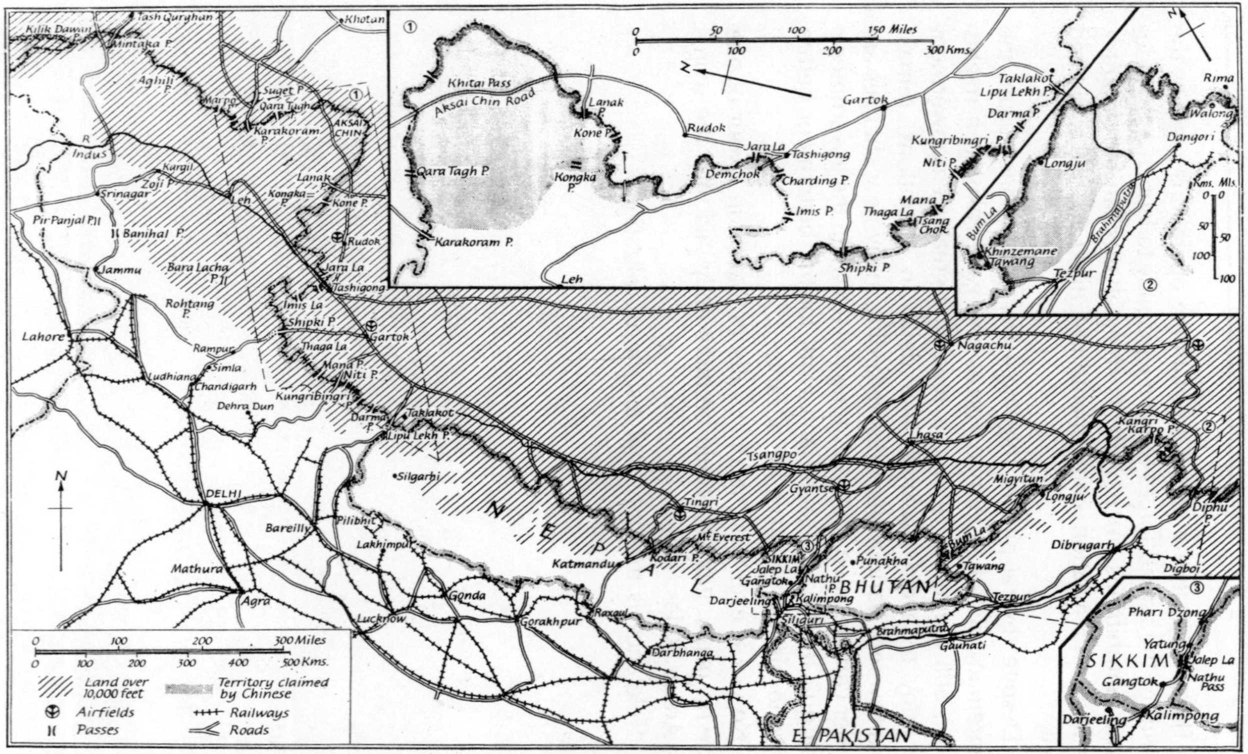

Indo-Chinese Border Area

(Map not to scale)

❮16❯

Epilogue

The unchangeable geography shows that most of the 5046 km border with China and northern border with Pakistan lies in mountains over 10,000 feet (2700 meters) shown as hatched. The unresolved border areas being discussed with the Chinese are also marked. Recently, the Chinese have accepted the Eastern sector from Bhutan to Myanmar, except only two details at Lonju and another place—a welcome sign.

They hold the Tibetan Plateau and control our trade routes to CIS States, rich in hydrocarbons. To ensure peace, we must present a strong military presence at our border. By 2020, most analysts project China with its fast growing economy as a super power.

The Lhasa Military Zone Commander commands the PLA which includes Air Force regiments, has access to a large fleet of transport planes, thus can switch troops. Working on internal lines with a railway line built to Lhasa, has few logistic problems.

Today our problems are enormous. Our supplies and ammunition have to be lugged up steep and scary roads. We have limited capacity vehicles. We require specially designed fuel-efficient vehicles, and a Chief of Defence Staff (CDS), to oversee co-ordinated defence by a C-in-C with adequate Air Force instead of control by four Army Commanders of—

Northern, Western, Central and Eastern commands, and an allocated Air Force.

The defenders, both holding and the strike formations, to be well equipped and to operate at these heights, backed by an efficient logistic system. Transport planes, helicopters and armoured planes like the U.S A-10s capable of providing close support to the Army. Ability to switch troops within the various sectors. Acclimatization of troops would be our only constraint, halts after every 1000 meters above 3000 meters altitude. To fight in these hills, we require lightly equipped infantry with effective weapons, clad to fight in the snows sustained by high-energy food. As a corollary, we have to reconsider our doctrine of mechanized warfare with heavy tanks.

Whereas, we luckily have many experts' tactical doctrines—the Armed forces officers' exposure to weapons production and development is limited. The Ministry having controlled it for the last 40 years.

The Hoover Report is relevant—it said:

"The management of defence establishments is no longer principally of managing tactical operations, of equal importance is development and production of weapons"

Weapons Production

Having talked of battles won and lost, the valour of our troops, and importance of their weapons, lets turn to their manufacture, and ponder over the problem of whether India can arm its defence forces adequately without hurting it's economic development—the theme of this book.

The weapons must be of high quality, excellent to provide the best fire power and raise morale, and in numbers consistent with the calculations as per Lancaster's Law[1], ensuring the theme of Efficiency, Timely supply and Economics.

Having been a subject nation for far too long, we accepted the wisdom of our founders, who ordained that the production of military equipments must be in Government factories. All Defence related issues had been dealt with by a "secret" department, outside the purview of the Indians. The same

"*Mantra*" was followed by the successor government even after Independence.

Our first Defence Minister, Sardar Baldev Singh, an industrialist, as head of a committee had recommended that all aspects of production and development of weapons and equipment including by civil industry should be under the unified control of a Controller General Of Production (CGDP). Mr. Wanchoo, the ICS, Joint Secretary in the Ministry, had headed the Ordnance factories as the DGOF as the first incumbent.

One day, I was called to a tea party by Krishna Menon (K.M), the Defence Minister, and asked to take over from Wanchoo. It was on September 27, 1957. Whilst returning after my first call, K.M who had known me since London days, walked out of his office to the corridor, put his hand on my shoulder, and said:

"Partap, you know I am dependent on Pt. Nehru, and as Dr Kothari has the ear of the P.M I would like to appoint him as the Scientific Adviser to Defence Minster." The appointment is still called S.A. to the Defence Minister.

K.M's first priority being publicity, he desired an exhibition of arms and equipment produced in Defence factories. This was arranged in Eastern Court. Amongst the exhibits was a copy of the Anti tank recoilless rifle which was placed behind a curtain.

My most interesting assignment was conducting Pandit Pant to the show. An astute person, he asked me if we paid the workers by results—I didn't know which way to look, the problem still dogs us.

For co-ordination of production, I had to tour extensively. Kohli, a crony of Jeep scandal days, was appointed as deputy CGDP. He helped me in disposing of the files.

Doctor Kothari was a gentleman, but his assistants were up to dirty tricks. Once they even claimed to have invented *Janata* radio receivers, one of which had been bought in the bazaar. I pointed it out to K.M. Unfazed, the Additional Secretary called a meeting on a holiday, October 8, and decided the Controller of Research and Development would work under the Scientific Adviser from January 1, 1958. The partitioning of Technical

Development Establishments started. Laboratories were established, away from production factories. Science graduates started being recruited.

On a visit to Ishapore rifle factory near Calcutta, I was shown the automatic rifle developed by Technical Development Officers and the skilled workers. I carried back two automatic rifles to Delhi and presented them to General Thimayya. During test firing, the scientists of R&D department questioned the killing power of the .30 bullet. This objection to its caliber, scotched its production. Finance would not even sanction the production of Jigs and fixtures to manufacture of prototypes. The fact that NATO forces had changed over to automatic rifles firing a .30 bullet and the Chinese had followed, was brushed aside. General S. D. Verma, the Chief of General Staff, failed to arrive at a caliber policy, he later resigned.

Another rifle, the light "Armalite," was shown to Gen. Thimayya and Brig B. D. Kapur, the CCRD, on their visit to the U.S. The fact that it too fired a .30 calibre bullet, was forgotten—it was backed by the R&D.

Americans sent a retired Brigadier to assist us with trials in India, my neighbour was their agent. The Ordnance factories examined and rejected it saying that the precision components would malfunction. The Infantry School Chief Pat Dunn told his deputy—my nephew—"We will drink their Scotch whiskey but will not approve their weapon!" In April 1962, there was an uproar in the Dutch Parliament on the rejection of Armalite rifles manufactured under license. By then, the manufacturers of F.N.30 rifle-demanded Rs 10 crores—as mentioned by General Kaul the CGS in his book *Untold Story*. The country went to 1962 war with single shot .303 bolt operated rifles.

Interestingly a Piffer, Lieut. Colonel Pritampal Singh studied the effect of the lack of automatic weapons in this war. He observed that the Dogras of 21 Brigade had used their LMG's effectively and made the Chinese run for the life. The LMG with an effective range of 600 yards, would have kept off the Chinese who were using AK 47's with range of 300 yards, but a Lt Colonel's voice was lost.

After the debacle, the Parliament insisted on the ouster of Krishna Menon despite Nehru's plea that he had contributed to defence production. (*Krishna Menon's role in harnessing science and technology must be downplayed. However I would say in his defence that he could not possibly be sponsoring an American weapon)*

T.T. Krishmnamachari took over as the Minister for Defence Co-ordination. His Secretary P.V.R. Rao phoned and called me back from Calcutta. Reaching Delhi, the Taxi driver at the railway station told me, "the Chinese have run away."

TTK gave us the go ahead and General R.G. Williams and his team finalized the design of the Ishapore self-loading rifle (ISLR).762–at last we had a better basic rifle.

Several years later, the ISLR was modified to a lighter weapon—the present INSAS.556-mm caliber rifle. Our rate of production could not keep up with the demand and 100,000 AK 47's were imported from Romania. The price of these—third lowest bidder was only $99 each the price of the $ was only Rs 17.03 in 1990. Which works out to Rs 1,686. Less than one tenth of Rs 17,000/-the price of our INSAS rifle.

Why is Cost High?

Basically because our factories work on cost plus 10% profit. There is lack of cost control—higher the cost higher the profit. When the new rifle factory was put up at Trichrapally by Vice Admiral Daya Shanker with the help of the United States, they offered two manufacturing processes. 'Barrel swaging' and 'high energy forming of the bolt.' Barrel Swaging would have saved 14 machining operations. High energy forming of the bolt would have eliminated 54 operations. The Ministry did not accept these hençe 400 milling machines were ordered on Bharat Fritz Werner.

The effect of installing the extra 400 machines, was employment of 400 additional workers plus '*badlis*' their wages, overheads—cost of additional factory space; largely explains our problem of high costs.

Even worse is the case of ammunition sold by ordnance factories @ Rs 20/-per round, whilst market price is Rs 6 only.

I was told because we use brass cartridge cases, whilst the rest of the world uses steel. Lately, we too have manufactured steel cartridge cases.

Some thirty years ago, we of the Reforms Commission, on a visit to Shahjehanpur clothing factory found it had no costing system. It was decided that it be closed but no action was taken as it interfered with the vote bank. Today, a jacket and a pair of trouser cost Rs 808, as compared to Rs 523 in the market—extra expense is Rs 285.

Recently, a friend and I visited a foundry that manufactures cast iron bombs and grenades. Foundries are murky places, but we found lawns instead. In the old days, it sent castings to Kanpur for machining. Now it uses electric melting furnace, CNC machines for turning and sends them by train to Khamaria for filling and finishing. It is run as a jobbing factory; it supplies casting for T 90 tanks to far away Avadi factory.

It has outpriced itself, the cost of castings is @ Rs 50/-per kg. No wonder the Indian Air Force imports bombs on their inventory.

Tank Story

K.M. was keen on indigenous production of tank. We started with manufacture of tractors. None of the Western manufacturers were willing to collaborate, so we asked Komatsu. The agreement was finalized between Admiral Shanker, myself and Komatsu, no one else was involved. I remember the contract was ready for signature about 7 p.m. and then the Ministry intervened to meddle with the prices. With growing demand for road making machinery, Bharat Earthmovers took over this lucrative business.

P. K. Basu of the ICS helped me to dispose of files, while I went to Tisco which produced armour plates and cast hulls in WW-II. We were lucky to be offered the German Leopard Tank for which the collaboration was to be through Mybach, a well known firm, our Ambassador Tyabji at Bonn helped. I visited their factories and was enthusiastic for formalisation of this deal till at a meeting with the Minister and Tyabji, K.M ruled that the

Government had decided to enter into an agreement with Vickers and project will be handed over to the R&D department. Remonstrations were of no avail.

Soon after I accompanied the Defence Minister on a visit to Avadi—a war time engineer stores depot which was to be closed. The Minister decided to set up a clothing factory due to nearness to Buckingham mills which supplied olive gray cloth.

Later, he also chose it to manufacture Tanks. It was far away from sources of materials or delivery. I did not sign the file and put in my papers.

When the first tanks were due to roll out just before the 1965 war, the two Vickers experts regretted their inability to align the sights and the gun till Payara Singh, a semi-literate supervisor, did it successfully. But none were sent.

Before the 1971 war, the Armoured Division found that these could not be used effectively as the tracks kept breaking. Vickers ruled that the alloy used in their manufacture was sub-standard.

The Arjun tank conceived in 1974 at a cost of Rupees 3 crores, is expected to be delivered in 2003 at a cost of Rupees 20 crores—80% of the value and the engine, and the fire control system are imported. Thus, its cost in FE is higher than the T90 tank which costs only 12.5 crores. Wither our Financial Advisers?

Costs of Wars

In June 2002, Sandip Dixit of the *"Hindu"* raised the question-how the 1965 war, cost us only Rupees 50 crores compared to the post Kargil bill for the Army purchases of over Rupees 3,000 crores? The management of Defence production had changed from Services to Secretariat control in 1963, when Admiral Daya Shanker the CGDP-retired at his request, but the benevolent system rolled on. Pakistani tanks were shot by CQMH Abdul Hamid from a Jeep mounted recoilless guns of our manufacture. Centurions of 3rd Cavalry fired Indian made APDS ammunition from the 20 pdr Guns to make a graveyard of 97 state of the art Patton tanks at Bikhiwind. Indigenous Gnat fighters downed the Pakistani missile firing Sabers. Pakistan had bought only HE shells which were unable to penetrate our armour. When the fact dawned on Ayub, he started sending messages for a ceasefire through the American Embassy.

Administrative Reforms Commission (ARC)

The loss of war in 1962 and the drawn war of 1965, led to a soul searching. The President appointed a Reforms Commission in 1966 to study the working of the machinery of the government, particularly the relationship between the Services and the Ministry. An integral part of it was a study group to look into Defence Research and Development, Ordnance Factories and other manufacturing undertakings. The Defence study team under the chairmanship of Shri Ravindra Verma, an M.P with a razor sharp intellect, late S. Moolgaonkar of Tatas, Kirpal Singh, Chairman, Railway Board, the author an ex CGDP, and Dr. Chawla, a scientist with experience of American aircraft industry as adviser visited numerous R&D, Production & Inspection establishments and also questioned the policy makers and submitted a report after three years of voluntary labour.

The report by luminaries like Moolgaonkar was thrown out, as it recommended that secretaries to the government should not be appointed chairmen of PSU's. Today no copy is available.

Two major reforms—appointment of a Chief of Defence Staff and establishment of a Defence Industrial College[2], to promote better understanding of Defence production problems at an intellectual level—still dog the country. The CDS appointment is in limbo, reportedly as the then IAF Chief objected, surely he could have been overruled.

Self Reliance in Aviation: In contravention of our recommendations, a Cabinet Secretary due to retire, hatched a scheme to carve out a secure job after his retirement. By simple maneuver—all the three companies—HAL, Aircraft India making Russian MIG planes, and the company assembling Transport planes at Kanpur—were amalgamated into one HAL. The corporate HQ of the company was moved to Bombay—the place selected for his retirement. It ensured the decline of Aircraft manufacture.

Between him and the Scientist Secretary, whose Doctorate was openly questioned in the "*Statesman*" as Director and Chairman of HAL, Aircraft industry had floundered. Presently, HAL in its present form is too large working on cost plus ten percent—it cannot but show profits, but needs a major restructuring. Whilst retaining it as a holding company for legal purposes, it must be split into specific cost centres with its HQ at Bangalore. It should have the center for R&D—Western Engines Manufacture—under it. The three centres should be as follows:

1. Nasik—with engine manufacturing plant at Koraput and Electronics at Wurwa, UP—for production of Soviet origin aircraft.
2. Kanpur—for manufacture of Transport Planes –utilizing Donnier 228 facilities, the Machining capacity at Ordnance factories at Kanpur, The HAL Electronics at Lucknow, the local industry and factory space of closed down ICI Fertiliser plant at Chakeri. Saras, the executive plane developed by the National Aeronautical Ltd. (NAL), could be at Hosur, the private factory of the Tanejas.
3. Kharagpur—or a site chosen by ISRO—the best organisation for getting co-operation from civil industry. For Light Aircraft UAVs+ Brahmos—or at Chandigarh, and Closing down the cable making factory, using the BRD, etc.

The HAL has been established at Bangalore—perhaps the helicopters and gun-ships may be left there and promote the factory manufacturing Saras, the executive plane.

Recently, the HAL has done an excellent job in producing and successfully test flying an intermediate jet trainer HT-36. All within a period of 2 years, it has a French Engine and a Russian ejection seat, a sensible decision.

A great pity is that whilst in the days of Air Marshall Malse, who on finding that the planes were grounded due to poor oil seals, had these analyzed and manufactured. Indian Air Force brass now only hold seminars on Air Power and Joint Operations. Recently, Air Marshall Mushraf Ali Mir of Pakistan hosted one on "Self Reliance" in military aviation. In his address, he talked of

Pakistan's manufacturing agreement with "Ketic" of China, which has led to induction of six K8 fighters, 24 are to follow. He also said that their Super-7, equivalent of our LCA, will fly by June 2004.

Imports

The production from our factories having lagged, we have become the largest importers of conventional weapons. Lure of commissions is compelling—I can do no better than quote what Air Chief Marshall P. C. Lal once told me. When as Chairman of the Indian Airlines, he entered into a very favourable contract for purchase of planes for the Airline. Happy at having secured a good deal, he went to report to the Minister. His comment is reproduced verbatim, "*Kiya Kar Ai Lal Sahib—na hamra bana na Apka*" (What a folly—Lal, neither I nor you have got anything out of it)

Imports of military hardware had reached the figure of Rs 62,000 crores in 1989. Considering that the dollar has since appreciated by 154%—that figure today would be Rs 94,460 crores. SIPRI report of the year 2001, reported our imports for the year 1998 were 4228 million dollars i.e. Rs 2.0,720 crores. No wonder, delegations from many countries flock to Delhi and throw lavish parties at their embassies for the DefExpo hosted by the MoD.

Research and Development—DRDO

A soldier perishes handling his tank, a sailor drowns when his ship sinks, an airman crashes to his death with his aircraft. The relationship between the combatant and his weapons symbolizes an intimate mutual dependence. After the exit of the CGDP organisation in August 1963, the development and production of weapons fell into the hands of scientists recruited in the DRDO, working directly under the Ministry.

The status built up by another S.A:, is best described by an extract from 1994 report by SIPRI (Stockholm International Peace Research Institute), an international and independent non-profit organisation, reproduced under:- The chief scientist "was free to make choices amongst projects—there was absolutely no oversight or control by any authority whatsoever." In the 10 years, he served under five Prime Ministers and over a dozen defence ministers and as a secretary with the longest service in the ministry, he was free to make unfettered decisions.

In 1987, Defence Secretary asked for advice on import of the Bofors Guns—the S.A., told him he was busy designing. They decided to import lock stock and barrel, literally. Thereupon, an Army Chief chipped in and asked them to cancel the contract to find where the Rs 64 crores commission went. This has resulted in non-transfer of technology for ammunition manufacture. We have been importing ammunition for 15 long years. Efforts for extradition of an Italian (Quattrochi) still burden the poor taxpayer, good money is going down the drain.

In the late 60s, facing a showdown with Pakistan, the Army Chief wanted advice on dealing with influence mines; the S.A. asked for Rs 10 crores for the project and 10 years time. A Professor at the IIT produced the solution in a couple of weeks—the Professor could not be honoured as DRDO objected.

Examples of DRDO being unable to meet requirement of the Services are a legion. I end with the Army's requirement of a close quarter weapon previously met by an invention B.C. Sen of the O.F's. called a Carbine. I learn they tried to make it by cutting off the length of the barrel; two reasons of rejection are—it got too hot, and noise level was 43 decibel. The result is expected in 3 years. An experienced manager at the H.E. factory said, "There is a lack of communications between user and designer".

Brahmos--Thanks to the legacy of expertise left by our scientist president, the DRDO has excelled itself by entering into a development and manufacturing contract with Mashinostroynia of Russia. It has successfully flight tested the Supersonic Cruise Missile with a maximum range of 280 km. With fire and forget operations, higher destructive capability aided by the large kinetic power of impact. It is expected to cost Rs 10 crores. It can be launched from land, sea or air platform, and has a huge export potential.

Tehelka

An Investigative attempt to find chinks in the defence procurement system was mounted last year which related to the sale of Night Vision devices, a project which has eluded our R&D for the past few decades. Image Intensifiers being a vital necessity during bad weather and darkness. The taxpayer is paying Rs 2 lakhs for each, considering the size of our huge land frontier most of which in the high Himalayas, we need a very deep pocket to meet our needs. The cost of import of thermal imaging sights for the Infantry was brought down as against the French quotation of 74,000 dollars each, Israel offered at 3900 dollars.

A more familiar device for looking into the dark is an Infrared scanner, as mounted on our tanks. These devices are based on detection of body heat emitted by an individual tank or a vehicle or a weapon after firing.

Sensors based on the Doppler effect sensing body movements, and microwave intrusion devices are sold by security alarm firms. Most important are electronic sensors manufactured by deposition of molecular layers of Gallium Arsenide, etc., in which several Indians are experts. Despite the right of Information bill passed by the Parliament in our closed society, it is not possible to find out if we are able to make them if so when? We have established a special laboratory GATEC. It reports production of some components.

The greatest setback due to Tehelka has been the negative attitude taken by the opposition against the Defence Minister George Fernandes who had undertaken 31 visits to Siachen, and cut the red tape for essential items.

Defence Production Secretary, during a recent seminar held at Confederation of Indian Industries (CII) seminar, is reported to have told them that in future tenders for such items will be displayed on the Ministry of Defence Website. A bold and forward-looking move. No tenders have yet appeared on the site.

The Navy

Late Vice Admiral Daya Shanker laid the foundations of Naval ship building when he initiated the establishment of a cadre of uniformed Naval Constructors. Lack of knowledge of oceans has deterred our scientists from interfering with the Naval planners. The Navy, which understands that "Navies are built not bought", has taken full advantage by acquiring the shipyards at Mazagon in Mumbai, and Garden Reach workshop at Kolkata in Krishna Menon's time. Despite the machinations of the import lobby which made it easier to import a ship than to build it, the Navy has been able to maintain steady progress in the building up of the fleet. Not only in numbers, it is their progressive modernization and better survivability which is the envy of others.

The new shipyard developed at Goa has recently built a 30-knot patrol boat, an example of its efficiency.

India's Navy is aiming to become a Blue Ocean force. Already familiar with nuclear powered submarines, it is poised to become the guardian of our nuclear deterrence.

The purchase of the aircraft carrier *Admiral Gorshkov*, the hull of which has been presented free and a 3-year lease of a *Akula II Class* nuclear submarine, has almost been finalised.

We really can't afford the luxury of an aircraft carrier—costs of maintaining one is billions and it only needs one torpedo or missile to disable it. A General, who participated at the proposed landing of our troops in conjunction with exercise "Brasstacs", told me that the whole of the naval fleet and its aircraft were entirely absorbed in protecting the aircraft carrier which was to give fire support. In any case, with the availability of Brahmos missile with its present range of 280 km and our policy of not coveting other people lands, the luxury of paying the cost not only of the *Gorshkov* refit plus its 28 Sukois does not appear to be justified.

The offer by the French to sell six Scorpion (SKK) design submarines fitted with a small nuclear reactor, has caused a complication. The French are masters of the art of salesmanship, they have already sold one Agosta 90 B submarine *P.N.S. Khalid* of the same design to Pakistan Navy. The answer to the question whether it is better to buy six SSK's at a cost of Rs 12,500 Crores or build 20 indigenously for Rs 15,000 crores by 2025, is obvious. Not withstanding Thales—the French electronic group—having already signed a transfer of technology agreement with Mazagon Docks.

Our Navy is familiar with Russian nuclear submarines, having operated one from December 1988 to 1991: offer of the *Akola II Class,* 10,700 ton nuclear submarine, fitted with 300 km range missiles, appears to be irresistible. Particularly, as the Russians would guide us with the design of the nuclear reactor for our Advances Technical Vessel ATVs on which we would need to spend $ one billion by 2007.

Navy's Engineer officers trained at *Lonavala*, and Electrical officers at *INS Valsura,* near Jamnagar, practical engineers with an Engineering degree, who keep their ships afloat are our greatest asset.

The Awakening

Luckily, at last, there are signs of change. After two years of consideration, the Government has granted six licenses to private firms to manufacture armaments in collaboration with foreign firms with equity holding of 26%. Five licenses have been granted to Larsen and Toubro for missiles, torpedos, small arms and ammunition, and one to Mahindra and Mahindra for armoured vehicles with an Israeli firm. Informal discussion with the industry indicates that they are interested in a very wide circle of manufacturing including aircraft.

Both are very well established firms—press reports mention that other firms like Ashok Leyland are also interested.

Co-operation from industry could greatly enhance our defence production base if the change is accepted gracefully. In the last decade, we put factory for armed vehicles at Medak to provide a safe seat for Indira Gandhi. Since then, fortune has smiled on Medak, another thousand acre factory has been put up by Bharat Dynamics for the manufacture of some 20-year old

Konkurs missiles, but no one knows what will happen to the factory when the recently improved missile goes out of service.

We also put up three Ordnance factories for manufacture of vehicle components at a cost of Rs 140 crores plus two gray iron foundries in this area, the sooner we hand these over the better. The Jabalpur factory produces the Shaktiman truck, a fuel guzzler, to a 40-year old design.

The production of ammunition is our greatest bugbear, year after year we surrender funds. Partly our high cost is due to use of brass cartridge cases instead of steel. The advantages of present ammunition running cool, less wear and tear to the barrel are minor. The speed of change over would set the pace. It is suggested that we follow the *American way of Government owned/operated factories.*

Thirty years ago as members of Reforms Commission with S. Moolgaonkar of Tatas, we thought of putting Kirkee factory in the hands of civil management. Pune is only two hours away by a super highway from Mumbai. The two High Explosives (HE) and Ammunition factory are distinct and separate. The HE factory's huge office in a separate enclosure could be the HQ of L&T Defence operations.

Leaving the HE factory with the DGOF, L&T could take over the Ammunition factory. It is a bit of a rabbit warren and would need to be re-laid out for easier access. The existing.556 ammunition production line, along with its stores, test equipment, the MoD Quality Control and security set-up remaining with the Government.

There would be problems with the transfer of redundant labour which could be solved. The costing of the assets would be eliminated as the estate, plant and machinery plus such other added for process improvement, will continue to be Government owned.

With a proper logistic system, whereby instead of the production orders are placed after determining the deficiencies at the unit level, we could work on the basis that small arms ammunitions has a shelf life of 15 to 20 years. For training

Infantrymen, 40 rounds and others including security 20 rounds are required per year. The production of the minimum annual requirement—divided amongst the three ammunition factories—would be the order on each factory. It would mean work of two to three months only for the L&T factory.

The industry can put the ammunition line on maintenance and use the labour for whatever else (within limits) it likes to produce.

This way we could arm our fighting services and remain economically viable.

All depends on the "Will to Achieve" and whether we can overcome the stigma of *moving forward in slow motion.* With a galloping budget deficit now running as high as 11% of the GDP, the Finance Minister has himself pleaded for better management.

Another problem is non-utilization of budget. Surrender of Army Budget for essential requirements of the troops was 29% in 2001 and 22% in 2002. The Government should allow servicemen in uniform to control expenditure as per their budget as in the British Defence Services.

Our complicated procurement system ridden with Financial Advisers (FA's), frightened of scandals, contribute to not equipping the Armed forces needs revamping.

All what the change in the procedure has achieved is addition of higher ranking civil and service officers. Service officers on their normal three year tenures will have little influence is getting weapons developed as required by their services.

Times of India, dated March 14, reported the scandal of import of 8588 sleeping bags from France at a cost of Rs 7.65 crores in 1992-93. For a country with hundreds of scientists employed on development of stores and four factories for production of clothing and equipment, it is a matter of great shame. Hopefully—some heads will fall!

On the same page was a report of Jaswant Singh, the Finance Minister, who once wore a uniform pleading before the *Rajya Sabha that with the interest payments and Defence burden he had nothing in hand.* How long can the Government wait to plug the holes in its finances? The Finance Minister has not done enough to drive the public sector. The budget has been burdened with the cash losses of the ordnance factories. Those days these could be treated as cost of war insurance. The Atom bomb having ushered a 90-day war the excuse is no longer valid. The Reforms Commission unable to find a solution had recommended another group to delve into the problem. It had also sought the closing down or sale of some. Yet, the Government added 13 more factories including three for vehicle components. No one remembered that Morarji Desai had only agreed to assembly of Shaktiman as Hindustan Motors's price of a spare rear axle was greater than that of the Bedford truck.

The latest news is that we shall have a lean and mean Air Force. The report that HAL has launched the licensed production of first 140 Sukoi's 30 and expects to hand them over by the end of 2004, is most reassuring.

Notes

1. Appendix IV
2. Appendix III

Appendix I

Aircraft—Gnats and Imports

It was on June 24, 1954 that I had accompanied the Air Chief Gibbs and Air Marshal Subroto Mukerjee, to the famous Farnborough Air Show, where a new plane the "Midge" made by Follands Aircraft, showed its paces. It was finally accepted by the Air Staff and manufactured in India, Powered by Orpheous engines, which we also manufactured, in collaboration with the Bristol Aircraft Company.

My association ripened when I was seconded to the RIFF. I found the site of Hindon Airbase, and worked on the development of Ambala, Agra, Pune, and Kalaikunda airfields.

Later as CGDP, I saw the development of HF-24. The induction of the "Gnat" the greatest, single engine fighter plane, nurtured and flown by the Indian Airforce was a landmark. It could climb to 40,000 feet in 4-5 minutes. This single engine plane, had such a low silhouette that it could come up behind the enemy, un-noticed, and in consequence ran circles round the Sabres of the Pakistan Air Force, in the 1965 Indo-Pak war.

Pakistan had purchased the F-104 missile firing planes from the Germans, with a nod from the USA. The Germans were glad to get rid of them as these were dangerous planes to fly, and were called "Widow Makers." Our pilots like the Keelor brothers faced the missile carrying Sabre jets of the Pakistan Air

Force, both in the 1965 and 1971 wars and downed them regularly. Phillip Nordoon, in his book on the Indo-Pak war, called the Gnats "Sabre Slayers."

Two other Air Forces, of Finland and Britain, flew the Gnat. The British made a two-seater version in a trainer role. Manufacture in India was a success story—80 per cent of the airframe and 60 per cent of the engine components were indigenous. It involved many snags; these were ironed out, with the close co-operation of the senior officers like Air Marshall C V Gole.

Unfortunately, we did not even follow up the manufacturing of the two-seater trainer.

Aircraft Imports

In 1956, we imported French Mysteres and by 1958, they were delivered, Israel at that point of time was also importing Mysteres. This was followed by import of 117 Canberra bombers and 200 Hawker Hunters. Whereas the French agreed to allow Israel to manufacture their plane, they refused us licensed production. America offered manufacturing rights of F-5 but withdrew the offer.

It was at this time when we asked the USSR for an engine for the HF-24 and were offered MIG 21. A license agreement followed.

The project was handled in a most deplorable manner. Three Ministers tore apart the project—Chavan located the Assembly plant at Nasik, Patnaik took the Engine plant to far away Orissa, Electronic went to Hyderabad. It was no way to produce planes.

Imports of the aircraft started with the purchase of Jaguars in 1978 by the then Janta Government. This aircraft with 2 engines was preferred to the MIG-23 with a single engine, which would have been produced at the existing MIG-21 factories. The decision taken, late one evening at Geneva, smelled of kickbacks.

This was closely followed by import of Mirage-2000 which flew about for 2 years without any armaments. The only disturbance to peace in the decade, 1979 to 1989, was our provocative exercise "Brass Tacks." Yet, it was the heyday for imports worth 17 billion dollars. These funds should have been ploughed into development of indigenous aircraft.

Today, we have only small aircraft manufacturer at Hosur, we don't use industrial power, and have been left behind in world technology. Even the simple UAVs, essential for reconnaissance and as target destinations, await imports. Remotely piloted vehicles for reconnaissance have been on the battlefield since 1982, when these were used by Israel. Our Defence Research and Development Organisation has been actively engaged in producing prototypes. A model named "Falcon" which was to be test flown in 1955.

Delegations have visited Israel and other countries. A few RPVs have been imported. The system has not resulted in production. These radio controlled toy planes are neither expensive nor difficult to manufacture. A Pakistani General had one produced by his ordnance factories within six months. These are now in regular use by the Pakistani forces on our borders.

These are a must for obtaining real-time information and for target designation. Their production is essential. A shame when a click of the mouse gives me details of 42 firms manufacturing these in America. China, whom we envy, has more than seven large aircraft manufacturing firms.

The important decision on upgrading of MIG-21 was delayed. The Russians had submitted a proposal for their upgrading—a similar proposal was also made by Israel Aircraft Industries. A contract for upgrading 125 IAF MIG-21 aircraft, with improved Radar and ability to fire advanced missiles, was signed with Russia in 1996. It took four years to clear the list of items required, the first two upgraded planes were delivered in the same year. Flight tests took place at Nasik on August 31, 2001. The programme is running three years behind schedule. The work has to be done at our own plant at Nasik. Meanwhile, every one is pressing for imports.

The frontispiece of the Defence Ministry report for 2002-03 was a MIG jet trainer. The HAL produced HT-36 in quick time using an imported engine. However with lot of pressure the government placed an order on UK for some 66 Hawk Advanced Jet Trainers at a cost of some 8000 crores. The trainers will only be available after 3 years, in the meanwhile some pilots will be sent to UK for training. The most worrying why a single engine trainer was purchased for flying twin engine fighters being inducted into the Indian Air Force.

Luckily, we now have Sukhois and HAL plans to manufacture these under license. The delivery from the first batch for 140 is promised by the end of 2004. The engine for these planes will be manufactured at Sunbeda in Orissa. Apparently we have not yet learnt that the engine and the plane must be manufactured at the same location. So that the aircraft can be sent to it for overhaul and repairs.

Transport Planes

The best we have is the Saras 14 seater designed by the NAL. China which had no industry at all is today well on its way to manufacture its 92- to 98-seater transport plane. Another factory in China is already assembling Embrar Brazilian planes which we now intend to import.

Appendix II

Tanks

In 1952, General Maharaj Rajinder Sinhji was the Army Chief, he persuaded the Government to purchase a Division's worth of Main Battle Tanks (M B T's). The British Centurian Tank with a 20-pounder gun ability to take punishment and tested in Korea was selected. As the Military Advisor at London I was intimately involved in the procurement and was the sole signatory of the contract.

These tanks proved their worth both in the 1965 and 1971 tank battles against the more sophisticated missile firing and night vision capable Patton tanks of Pakistan.

The situation in the year 1965 was extremely precarious, when the Pakistan armour broke through at Khemkaran, it could have made it to the Beas River. The Army Commander Lieut. General Harbaksh Singh had confidence in his troops: the Sappers flooded the area, planted antitank mines, and he held.

We were successful in destroying the superior Patton tanks because of the excellent training of our antitank gunners. In 1965, some Pakistan tanks had run only for 25 kilometer and their crews were not trained to operate computer-controlled tanks. Further, as these were driven by gasoline engines, they were liable to catch fire.

The result of the battle was a graveyard of 97 Pakistani Patton tanks at Bikhiwind, thanks to the 20-pounder guns on the Centurions, firing AP Shot, and our other stout defensive steps. Havildar Abdul Hamid with his Indian produced RCL (Recoilless Anti-Tank Gun) killed the Pakistan artillery Brigadier's tank.

In both the wars, officers of the Poona Horse, the regiment equipped with Centurions, won the highest bravery decorations—Param Vir Chakras, for foiling Pakistan armour. It earned the title of "Fakhar-a-Hind" from the enemy.

Despite their excellent performance these tanks were disposed off to South Africa by a relative of a Prime Minister. The South Africans, like the Israelis, fitted new engines to the perfectly sound chassis and used these disposal tanks as their Main Battle Tanks.

We had been in quest of an indigenous MBT .The proposal to manufacture the German Leopard tank was not approved by Krishna Menon the Defence Minister, who decided to manufacture Vijayanta tank in collaboration with Vickers of England.

To finalize the agreement two delegations were sent abroad. The first delegation was headed by Lt General L P Sen. The second delegation in mid-1961 was headed by Mr. Sarathi, the Additional Secretary, with the financial advisor, and signed the development contract with the firm. The design was on the drawing board only. Later, the Director of Armoured Corps Maj. General K C Khanna found that the tank did not even have a coaxial machine gun. This omission was rectified in 1962 at an additional cost.

Colonel Samson took charge of the factory and headed it for eight years. When the first tanks were due to roll out just before the 1965 war, the two Vickers experts regretted their inability to align the sights and the gun till Payara Singh a supervisor did it successfully.

Production however was held up for want of the imported 105 mm guns, transmission, and engines. Therefore 40 tanks were imported from Vickers and 620 tanks (T56 and PT74) from Poland and Russia.

Interestingly Vickers also sold the same design to the Shah of Iran and two expensive computerised simulators to us for training of tank crews.

The Arjun tank was conceived in 1974. In view of delays in manufacturing and approval, as it was too heavy, and no sign of manufacture of the Arjun tank in development stage, the army imported T72 Russian tanks which have been the main battle tank for the last 20 years, and these have now been replaced by the T90s.

Presently 124 Arjun tanks are on order for production at Avadi. These are expected to be delivered at a cost of Rupees twenty crores—80 percent of the value—the engine, and the fire control system are imported. Thus its cost in foreign exchange is higher than the T90 tank which costs 12.5 crones.

Hard facts are that the Arjun tank which was so called accepted by the Army in 1953 has not yet been delivered. Delivery of 6 tanks is expected shortly. In the meanwhile with the development of UAVs and their capability of directing aimed missiles at comparatively slow moving expensive tanks, their survival in the battlefield is doubtful.

Appendix III

Defence University—Industrial College

It is said that at the time of Dunkirk there were three British P.Ms near the beaches. An American president was accused of dodging the draft; the U.S. has a Defence University where the art of war and the means of fighting war, are debated.

We with neither veteran PMs, nor a Defence University: mobilized on December14, 2001, the day after attack on our Parliament –but took five months to implement the cardinal principle of "Unity of command." Only on May 14, 2002, when Army wives and children had been killed in an attack at Kaluchak.

The seven principles of war summerised, as "Moses CC" must also be imbibed by our leaders, and their civilian advisers, who guide and demand the country of taxes.

No educated military person will deny that he is answerable to the political masters, whose prerogative is, to use war as an extension of their diplomacy. Surely, it is imperative that we too need to set up a Defence University for the benefit of our future leadership. The Government has already accepted this fact in principle.

A splendid opportunity of building such a University Campus; can at the village for the Commonwealth Games in 2010. I suggest the Prime Minister, should also remove the stigma based on the remark by George Tanham of the Rand Corporation—"traditionally Indians do little formal strategic thinking."

The dismal handling of 'Operation Parakaram' proves that it brooks no delay, as we failed to honour the first principle of war, namely "Maintenance of Aim." In the case of Op Prakaram –{Infact the aim of the war was never printed-?}

I quote from page 62 of Lieut. General V.K. Sood ex Army Vice Chief's book co-authored with Parvin Sawhney, published by SAGE Publications. On Dec 18th on being asked by General Padmanathan the Army Chief, "what the Government expected from the war." The P.M. told him "*Woh bad main bataayenge*" (that will be told later).

Yet in the Parliament the P.M. on December 14th 2001, had thundered of an *Arr Paar ki ladayee* (a war to the finish) (p94).

Our two demands were Pakistan should stop cross border terrorism, and by the Home Minister that 20 criminals in Pakistan should be handed over. None of the 20 criminals were handed over.

After the attack on Army families at Kaluchack on May 14, 2002, the P.M. visited Kupwara on May 22, and told the soldiers to prepare for a decisive battle.

At the end of the ten months of mobilization at a cost of 8,300 crores and 130 lives lost on laying mines on the border, we ordered "strategic relocation" and sent the army back from its battle positions.

The Defence Minister told the Lok Sabha that infiltration was 53% of the previous year.

General Musharraf on the same day declared that the honour and dignity of Pakistan was upheld by not succumbing to the Indian pressure.

We thus, ordered mobilization and kept our troops out at their battle positions without much gain. Pakistan ruled by dictators has the initiative for action—ability to "surprise." We lack it, as a democracy. Yet, we are not the only democracy, others preserve their independence.

Defence Industrial College-as a part of the university, too has been approved by the Government and awaits implementation.

It is essential today to stop the craze to import; it is imperative to draw in Industry, Academics and the users together to innovate. Basically, we have a blinked approach on the role of technology in future wars. Confusion starts at the role of science. I knew the great showman Sir S.S. Bhatnager, who impressed Lord Irwin with his demonstration of a duck sinking in water. He started to build a chain of CSIR laboratories. A worthy Scientific Advisor to the Defence Minister, copied in fact surpassed him by building 52 Laboratories.He too started with metallurgy, chemical –to food and Electronics etc for the Defence; as if the food for a person employed on defence requires a different science!

All of us appreciate the importance of providing adequate weapon systems in the field that decide the fate of nations. For taking advantage of technological improvements—it is important to cultivate a triumvirate –The intellectuals, Professors at Universities with the industrialists and the military men and women.

Development and production of the means of waging wars – are intertwined. A fact that has been well appreciated by Britain. The result is that both Oxford and Cambridge, and other Universities are extolling their efforts in furthering the development of technology.

We normally follow the British pattern –unfortunately we have not been keeping up with their evaluation of thoughts on Defence. There is a sea change in the military education field. I have before me, the brochure, on the Royal Military College of Science—the premier institution for training future technologists of the services. The Cranfield University runs the institution, and the Dean who looks after the education and syllabus of the Royal Military College of Science is our countryman from Ludhiana.

The Defence Industrial College will be a meeting place between the nation's military and industrial arms. Cooperation between the two facilitates – almost insures that the Armed Services will have new equipment. The procurement of adequate equipment is central to the capability of the Armed forces fighting a war-thus it is a national necessity.

Late S. Moolgaonkar, of Tata industries, and we members of the Study team of the Reforms Commission for three years, had recommended that for a wider look in matters of Defence Production there is an imperative need of a Defence Industrial College, to expand at relentless speed, in various directions-beyond an individual comprehension. Its utilization requires interaction besides providing a forum for partnership with the private industry: it would give direction to the development of a cadre of industrial engineers in the country as an essential requirement for job creation.

1. It would also serve as a "Think Tank" for development of Industry and in way render guidance similar to the MITI in Japan.
2. A certain amount of progress has been made in involving industry. The Defence production department of the Ministry of Defence offers Rs 50,000 prizes, mainly to small-scale industries; much more is required.

   Striving for excellence is only possible by open competition.

   Today Electronics is a major component, 30-35 % of all military hardware. Minds can only be opened for wider outlooks-by university education.
3. The Government's stress on Swadeshi is perfect. But a country that can not produce a trainer for its Air Force has no business to talk of military prowess. At the end of the Second World War, outside the western world South Africa, India, Egypt and Brazil had viable defence industries. It is also on record—in a paper by Beniot—the economist, that India's economy improved in 1960-61 because of its defence production. Those were the years when the HAL produced the famous Gnat fighters-the Sabre slayers. After that the whole system was covered up in a veil of secrecy. India became the top importer.

Defence Industrial College is essential to formalise attempts already being made by our intellectuals—towards a closer cooperation with the industry. Meetings are held by our Scientists and Engineers with the CII—Chambers of Indian Industry. There is need of a formal venue for interaction at the highest level between the Captains of Industry, Bureaucrats, Generals, Air Marshals, Admirals, Scientists and Engineers.

Our leaders need exposure to advanced technologies. It is they who travel, visit industries throughout the world. Through them we would enter the gates of dual usage technologies. They would acquire them to expand their businesses, and as a bonus help defence production technologies. For example, opening of the world of sensors, the basic component of surveillance and precision guided bombs would come through, development of usage of Gallium Arsenide—a byproduct of our Aluminum industry. We have to work together, let's try to learn from others.

A way used by the Industrial College of the U.S. –for dissemination of technology –is that, each student selects one type of industry and then for the dissemination of technology sponsors, at least, one speaker for advanced technologies of his or her choice.

Each student selects one type of Industry in which he or she may be specializing:-

Automated manufacturing systems, Electronics, Armaments, Landvehicles, Ships, Aircraft, Fossil Fuels/petrochemicals, Space, Telecom etc.

They, then undertake a field study of domestic and international industries, their capabilities under normal and emergency conditions. Suggest changes in policy.

Students also undertake special survey trips—one third each to Far East, China and Russia. I met the students visiting MDL –Mazagon Docks Ltd.

In the meanwhile, under the patronage of Ministry of Defence, an association (CDOT) of defence undertakings, in the public sector including DRDO has been set up. A well-published seminar of this organisation was held in January 2003, in conjunction with the USI (United Service Institution). A good number of servicemen attended. Though this organisation is useful from the point of view of exchange of information between the various defence industries, it is only complimentary to Defence Industrial College, as the heads of the various production units themselves are not technologists but are administrators.

The requirement of a Defence Industrial College is irrefutable and cannot be delayed—The latest news that the engine for the Sukoi planes well be manufactured at a 320 crore factory being established at Sunbeda in Orissa state is the last straw.

Surely, by now, we know that the plane and its engine must be manufactured at the same location–so that the planes can be sent there for overhaul and repairs.

Appendix IV

The Lancaster's Law

Lancaster's—Inverse Square law is best explained by an example. A machine gun is 16 times as effective than a rifle. But the number of machine gunners to replace a 1,000 men battalion of riflemen is not 1,000/16=62.5.

Due to other factors such as the need to hold ground, etc. it will not be possible for a battalion of only 62.5 riflemen armed with MG's to replace them.

The correct number of men required for the task is arrived at by dividing 1000 by the under-root of 16—which is 4, i.e., 250.

This equation would be written as under:

1000/divided by under-root of 16 (i.e. 4)=250

QED—A battalion quipped with a weapon 16 times as effective would need 250 men.

A weapon twice as efficient would need 710 men.

For small advantages in performance, the reduction in the manpower is infinitesimal, a fact generally ignored by purchasing authorities. They are apt, to be swayed by the sales talk of the vendors.

QED—Sales talk of superiority of imported weapons, generally, is not worth the extra cost.

Appendix V

Outline of New Procurement System

The recent reform has produced the following system for procurement:-

1. Defence Acquisition Council—under the D.M. Members RRM, CDS, Chiefs of Staff, VCDS, and four Secretaries –viz Def, Def DDPS, Def R&D, Special Acquisition and the F.A (eleven) will approve long-term plans. Identify make or buy projects and monitor the three boards viz:-
   - A. Defence-Procurement Board –under Def. Secy, Spl. Secy, will have Vice Chiefs, VCDS, FA, and other Secretaries as members.
   - B. Defence Production Board. Chairman, Secy. Defence Production and Supply.
   - C. Defence R&D board-Chairman, Secretary Defence R & D .
2. Each Land, Marine and Air—will have a Joint Secretary as Acquisition Manager and a two star service Technical officer.

   It is noticeable, that even the council which is to "make or buy" decision has no Production Engineer Adviser. Another stipulation is that projects costing over 200 crores, a Project Management Organisation (PMO) should be set up. For other projects, a user representative can be co-located with the Development Agency. Yet another stipulation to keep out the users is that "The officers posted to these PMOs should have a longer tenure of up to 5 years."

3. A defence Technical Council-DTC-under the D.M. is to assist the Defence Acquisition Council .

The Defence Acquisition Wing-under the Special Secretary is to assist "A" above.

Exports-The export of defence material should rightly be under the procurement executive who is in touch with the world's industry, and therefore offers a single window to other countries and is in a much better position not only to obtain order but to insure deletions of component or even subassemblies from imports

The main flaws in the system as now installed, are recounted hereunder:-

1. The procurement system is not oriented towards manufacture of the weapons. The procurement executives should have a background of engineering. They must be assisted by industrial engineers, to at least pick out the components, available indigenously, which need not be imported. Under the Australian system the local industry, offers to supply such components, assemblies to the exporter as a counter balance to reduce the import cost.
2. Being a purchase system based on price alone: it overlooks the need for standardisation. The result is particularly harmful for the Airforce with planes purchased from different countries, each with its own standards. The multifarious types present problems in deployment—their maintenance is nightmarish.
3. Adhoc purchases from varying sources leads to difficulty in supply of spares. Without genuine spares availability, and repair in the battlefield, weapons turn into liabilities. For weapon platforms like aircraft, tanks etc, the only

effective way to overhaul is to send it to the manufacturing factory. Hence straight purchases without setting up manufacturing is self defeating .

4. The setting up of three separate boards each for land, sea and air is a very expensive way of working. In the fifties K. B. Rao as head of the DGS &D in London, purchased for everyone. The British and French have only one head of procurement.
5. The final step of Indian purchase procedure involves bargaining for the 10 % extra profit. The service representative is a silent spectator. There are chances of corruption; it may be worth while following the British system of making the executive have a session with a physiologist

The Reforms Commission, recommended that the Secretaries to the Government, should not be appointed chairman of PSU's, because it is not correct for a person to oversee –the working of an organisation run by him. It will be noted, that the Chairman of the R&D board is the Scientific Adviser running the department. Thereby there is no fresh look into the problems.

The present control system is unsustainable. I quote example of the anti-aircraft missile Akash, as reported in the press July 26, 2003. A DRDO scientist, told the reporter that Akash an anti-aircraft missile could be turned into an anti-missile. It is a far fetched wish – the Akash is yet to become effective against aircraft!

Index